THE OVARIAN CYCLE OF
MAMMALS

UNIVERSITY REVIEWS IN BIOLOGY

General Editor: J. E. TREHERNE

Advisory Editors: T. WEIS-FOGH

M. J. WELLS   Sir VINCENT WIGGLESWORTH, F.R.S.

ALREADY PUBLISHED

OTHER VOLUMES ARE IN PREPARATION

THE OVARIAN CYCLE
OF MAMMALS

JOHN S. PERRY

D.Sc., F.I.Biol.

HAFNER PUBLISHING COMPANY
NEW YORK

1972

599.016
P 463

In the U.S.A.
HAFNER PUBLISHING COMPANY, INC.
866 Third Avenue
New York, N.Y. 10022

OLIVER & BOYD
Tweeddale Court
Edinburgh EHI IYL
A Division of Longman Group Limited
ISBN 0 05 002343 8 Hardback
ISBN 0 05 002342 X Paperback

First Published 1971
© 1971 John S. Perry

Printed in Great Britain by
BELL & BAIN LTD., GLASGOW

Preface

The aim of this book is to provide for the student an account of the mammalian reproductive cycle, centring on the ovarian changes. The intention is to present what is known in concise form, and to indicate the directions in which research is moving. Simple as it may appear, this is actually a formidable task. It is always difficult to be both concise and accurate, and doubly so when, in trying to cover recent work, one is hedged about with reservations. However, the intention is good and clear; such a synthesis, explaining the basic physiology and describing its modifications in laboratory, farm and wild animals, and their exploration by experimental and ecological methods, has not hitherto been attempted as far as I know. But does this mean that the attempt is foolhardy? Suppose, for example, that the problem of 'luteolysin' were suddenly to be resolved, much current discussion would be rendered obsolete, and what now appears to be central to the problems of reproductive physiology might be relegated to a by-way of history. Risk of this sort has been accepted in the hope that the discussion of live issues may engage the student's attention, but with consciousness of the risks of confusion.

The arrangement of the book has been changed from my original plan. After describing the developmental and morphological background, I at first proposed to describe the variations as they occur in nature, and then to deal with the physiology behind these phenomena, but I have reversed the second and third of these components. In this way the physiology, as studied in the laboratory, is described before we consider its manifestations in the field.

No component of the reproductive cycle can be considered in isolation, and I have tried to minimise the repetition that this fact engenders by the free use of cross-references within the text. As to references to the literature, I have tried to take the facts from the classic researches of the past and from the most recent sources available to me. In the former case (and sometimes in the latter) I have brought the author's name into the text. I have not hesitated to quote directly a passage in which some point is felicitously expressed, or when an observation seems to be most accurately conveyed by using the words in which it was first reported.

The list of references is intended to quote authority for most of the statements made in the text and, perhaps more importantly, to provide further reading—a 'way into' the literature in various areas in this huge field. To this end, references to the more general accounts, as well as to some of the milestones of past progress, are marked (*) in the bibliography.

I have been, and am, very fortunate in that the Institute in which I work possesses a superb library, and I am greatly indebted to the librarians for their most helpful efficiency. For much of the information and ideas I am also indebted to my colleagues, and for practical help to Mrs Nicola Ackland, who prepared the text-figures. Plates II and IV are from electronmicrographs by Miss P. R. Crombie, and Plate V is from photomicrographs lent by Dr. I. W. Rowlands.

JOHN S. PERRY

Contents

1: Introduction

The principal land-living vertebrates, the reptiles, birds and mammals, are characterized by the possession of an amnion, enclosing a fluid-filled cavity that provides the embryo with an aquatic environment—its own 'private pond'. The amniotic fluid provides a medium in which the embryo can grow and differentiate, and develop limbs, in a condition of virtual weightlessness, in that it is suspended in fluid and therefore subjected to equal pressure over its whole external surface. The amnion itself does not, however, 'protect' the embryo; it needs to be enclosed and supported. The sauropsid solution to this problem is the shelled egg of the reptiles and birds: the mammalian solution is viviparity.

This divergence is one of the most striking features of the phylogeny of the Amniota, and it is accompanied by a corresponding divergence in ovarian function. For the sauropsid egg is typically supplied with a large amount of yolk, and the deposition of this yolk within the oocyte is the predominant function of the ovarian follicle in which the egg develops. In the mammals, the nutritional function of the ovarian follicle is very greatly reduced and its hormonal function is correspondingly enhanced.

The cleidoic egg is not confined to the sauropsids, and viviparity is not the sole prerogative of the mammals,[6] but the mammals, particularly the eutherians, are unique in the degree of development of ovarian hormonal activity and in the extent to which the ovarian cycle is modified to accommodate pregnancy.

Mammals are first and foremost animals that suckle their young, although the monotremes are oviparous and initially provide for the sustenance of the embryo by a (relatively) large yolk. Even in this group, the young, when hatched, are fed on milk secreted by their mother's mammary glands. The marsupials, in contrast, have a nearly yolkless egg and intra-uterine placentation. Their reproductive cycle is fundamentally similar to that of the eutherian orders, but it differs in one very

1

important respect—placentation is brief and the rhythm of the ovarian cycle is not modified by pregnancy. In the eutherian mammals the ovary is involved in the maintenance of pregnancy and the ovarian cycle takes on a new complexity. In the chapters that follow, the monotremes, and even the marsupials, will be dismissed rather summarily. It must be admitted that this is at least partially because of the paucity of information, especially with regard to the monotremes, but it is the eutherians that provide the fascinating range of variation among wild species, the economically important characteristics of the domesticated animals and the endocrine mechanisms that point to the elucidation of our own physiology.

Vertebrate reproduction is essentially cyclical in nature. Its periodicity is determined by the influence of external factors upon the individual mechanism, and mammalian reproductive processes are characterized by a basic uniformity underlying a remarkable diversity. The ovarian cycle is initiated by the action of pituitary hormones (gonadotrophins) and is governed, after puberty, by an interplay between the ovary and the anterior pituitary gland, with the participation of the uterus and, in pregnancy, that of the placenta. The hypothalamus is involved in the control of the pituitary gland; it exercises an integrative function, not merely relaying but coordinating impulses from other organs and from the outside world. The multiplicity of routes by which the participating organs can affect each other, and the differing emphasis placed on the role of each in different species, gives rise to the variety of what Asdell so aptly termed the 'patterns' of mammalian reproduction.[20]

At this point, it may be as well to emphasize that although the more striking features of mammalian reproduction are rhythmic, in some essential respects it is non-cyclic. The rhythm is imposed on the ovary from without, and is not initiated until the ovary, in the course of its development, has acquired the capacity to respond to the humoral agents which thenceforth control it. The primary 'product' of the ovary is a stock of oocytes; these develop within the ovary, their production is non-cyclic and is not, apparently, under pituitary control. It is not the formation but the release of oocytes that is cyclic.

In planning a book on this subject, the main problem is to organize the available information in some more or less logical sequence. No matter what headings one selects, the discussion of each topic seems inevitably to involve almost all the others, and the dangers of being repetitious appear unavoidable. Reproductive physiology is in fact a study of integrating mechanisms. The words 'harmony', 'balance', 'integration' recur in the literature, chronologically displacing each other in that order as the complexity of the processes is revealed, and as the techniques of investigation are increasingly refined. The inter-relationships between the participating mechanisms, and between the various factors that modulate their functions, are very forcibly borne in on the author in this attempt. It is to be hoped that it is conveyed to the reader as an impression of harmony and not one of confusion. The harmony within the systems involved is real, but there is also confusion in the literature concerning it. It is my ambitious hope to describe the former without adding to the latter.

The cycle in monotremes

The mammals evolved from the Therapsida, an order of reptiles which had for long been distinct from the sauropsid branch that gave rise to the birds. The earliest mammals are, of course, known only from the fossil record, which does not tell us such things as whether, for instance, they were homoiothermic. It would seem probable, however, that their mode of reproduction was similar to that of the reptiles, since the egg-laying habit survives among the most primitive of the extant mammalian species, the duck-billed platypus *Ornithorhynchus anatinus* and the five species of *Echidna*, the spiny ant-eaters. The platypus and the echidnas are the only living representatives of the sub-class Monotremata. In other respects they are indubitably mammalian; the simple but fundamental definition of a mammal as 'an animal that suckles its young' holds good for them, even though the young are hatched, after incubation, out of a shelled egg. The egg is small by avian standards but of a different order of size from that of other mammals. The egg of the platypus is similar in size and shape to a thrush egg; the echidna egg is spherical and about 4 mm in diameter. The 'shell' is leathery in texture, resembling the shell of a reptile's rather than a bird's egg.[186] The echidnas, like the marsupials, have a brood-pouch, but it regresses each year, after the breeding season. The egg, when laid, is placed in the brood pouch by the mother, a procedure which is aided by the fact that the cloaca is protruded in such a way that its opening is brought very near to the pouch. The egg is incubated in the pouch, and the young are retained there after hatching. The eggs of the platypus are incubated in a nest.

The ovarian follicle of the monotremes, like that of all mammals, is transformed into a corpus luteum after ovulation. This structure is not phylogenetically 'new', for a corpus luteum is formed in a similar way in the ovaries of many reptiles. It is not known whether the corpus luteum has any endocrine function in the reptiles or in the monotremes, but in the latter the egg does pause in its passage through the uterus, and the uterine glands appear to secrete a nutritive substance which is absorbed by the oocyte for some time before the shell is acquired. The mammalian trend is further foreshadowed in the monotremes in that, although the ovary is like that of the reptiles and birds in its morphology, the follicular epithelium reaches its maximum development at the time of ovulation, as in the marsupial and eutherian mammals. In the reptiles and birds the follicular epithelium is most active during the period of yolk accumulation, and regresses before the egg is released. Immediately before ovulation the follicle wall secretes a fluid around the oocyte; this is homologous with the fluid that fills the antrum in the graafian follicles of the marsupials and eutherians. Slight as these similarities may seem, they are probably significant. The structure formed in the ruptured follicle of the monotremes after ovulation, like that of the reptiles, is properly called a 'corpus luteum' because this is a purely descriptive morphological term, but it would seem that the mammals possessed the structure for some time before they, so to speak, found a use for it. By analogy with what is known of the corpus luteum of some reptiles we may expect that of the monotremes to elaborate some steroids, probably including progesterone. It would seem, therefore, that in developing their special type of viviparity, the mammals found a use for steroids that were already available, but we have no means of knowing at what stage of evolution, or in what manner, ovarian progestagens became functional.

The monotremes do not appear to have been studied in recent years, at least with regard to their reproductive physiology, probably because of their rarity*. The growth of the follicle and oocyte, and the formation and possible function of the corpus luteum, were mainly described from material collected many years ago.[134, 145, 185, 187]

The cycle in marsupials

The marsupials have a small-yolked egg which is ovulated from a

*But see *Echidnas* by M. Griffiths. Pergamon Press, 1968.

'mammalian' type of ovary.[115] The corpus luteum is functional as in Eutheria and the embryo is retained for a time in the uterus, where it is nourished and supplied with oxygen through an allantochorionic placenta. The ovary and the accessory reproductive organs of the marsupials appear to function similarly to those of the eutherians, but gestation is completed within the normal ovulation interval. A minor exception to this rule is the swamp wallaby, *Protemnodon bicolor*, in which ovulation occurs a few days before parturition, instead of a few days after, as in most of the sub-class. The normal cycle always includes a luteal phase, and the interval between ovulations is commonly about a month.

Although marsupial young are very small at birth, they cannot truly be said to be extremely immature. The condition of the neonatal animal varies enormously among the eutherians, and one could argue that the nest young of the rat are more helpless than the neonatal kangaroo, which travels 'under its own steam' over the surface of its mother's belly from the vulva to the pouch. It weighs, however, only a minute fraction of its mother's body weight, and it is evident that differentiation and organogenesis have proceeded rapidly in comparison with the total amount of tissue growth. By the time it is weaned, the young kangaroo is similar in size, and general independence, to the weanling of an eutherian herbivore of comparable adult size. The outstanding characteristic of the marsupials is that during the period of dependence between conception and weaning, lactation has a decidedly more important role than placentation. Australian zoologists have studied the marsupial reproductive cycle with great success during the past 15 years or so.[344, 345, 374] Their work is the more valuable and interesting because they have been able to combine ecological and physiological investigations, studying the animals both in the wild state and in captivity, observing behaviour, administering hormones, and comparing a variety of species from different habitats. In some respects the marsupium replaces the womb, but of course the infant has to be air-breathing during pouch life, it has to be mobile to some extent, and it has to absorb nutriment through its own gut wall.

In 1964 Sharman and Pilton described the extraordinary process of birth and pouch-entry in the red kangaroo.[345] The weight of the young at birth is about 750 mg, whereas its mother weighs about 27 kg; this is a ratio of about 1 : 36,000; in man the corresponding ratio is about 1 : 14, and in some eutherians it approaches 1 : 1. Other marsupials are even smaller than the kangaroo at birth, and some species of *Dasyurus* (the 'marsupial cat') weigh only 12·5 mg.[188]

Whereas pregnancy has little or no effect on the ovarian cycle in marsupials, lactation has a very marked effect. Oestrus occurs after its usual interval, very soon after parturition, as described above, but it is followed by a lactation anoestrus. This is common among eutherians too (see p. 74) and it will be seen that here, as so often in a study of biological adaptations, we are dealing with differences in emphasis rather than fundamentally different mechanisms. Again as in many eutherians, fertile mating may occur at the post-partum oestrus so that an embryo, in the morula stage, enters the uterus very soon after the new-born young enters the pouch. It will not be born, however, until after the pouch young is weaned and leaves the pouch, for lactation not only inhibits ovulation but also causes a delay in implantation of the embryo in the uterus. A second pregnancy may, therefore, be much longer than the first, and much longer than the oestrous cycle, both ovarian activity and embryonic development having been arrested, as it were, by lactation. This phenomenon parallels the 'delayed implantation' which occurs in some eutherians when eggs are fertilized at a postpartum oestrus so that pregnancy is concurrent with lactation (p. 118). The gestation period of the eutherians, however, is normally longer than the ovulation interval even when implantation is not 'delayed', and irrespective of lactation; in them, in contrast to the marsupials, the ovarian cycle is modified by the presence in the uterus of a functioning placenta.

The young ('joey') of the red kangaroo may remain in the pouch for about 7 months, before emerging for a short time. After this, its excursions abroad become longer and more frequent, but it does not leave the pouch for good until several weeks after its first emergence. If the pouch young is removed prematurely, however, the embryo which has been lying free in the uterus implants, and it is born 31–32 days after the pouch is emptied. In a related species, the grey kangaroo, which is one of the largest marsupials, mating does not normally occur immediately post-partum, but only after the pouch has been vacated. In this species, lactation is associated with a delay in oestrus and ovulation, but as there is no embryo in the uterus during lactation, there is no delayed implantation. Since the joey remains in the pouch for about 10 months, and pregnancy lasts about one month, the grey kangaroo normally has one offspring each year. If the joey is prematurely removed from the pouch, the female enters oestrus within 13 days.[303]

The similarities between marsupial and eutherian reproduction are decidedly more fundamental than their differences, but the extension of the luteal phase to accommodate a long gestation period must have been of prime importance in mammalian evolution. To the endocrinologist it is the central phenomenon of eutherian reproduction. It made possible a wide range of variations, and it appears to provide a more efficient relationship between mother and offspring. One is tempted to think that it is because of the relative inefficiency of their reproduction that the marsupials have not colonised so wide a range of habitats, nor spread over so great an area of the globe as the eutherians, but the assessment of evolutionary 'success' is necessarily speculative and probably misleading. There are examples, among the older groups of animals, of the survival of what seem to be extremely improbable adaptations, and the extinction of others may have been due, in reality, to the failure of the species from causes other than those which we can perceive or imagine. Furthermore, in the case of the marsupial habit, there are extraordinary examples of 'parallelism' between eutherian and marsupial genera (e.g. the marsupial mole, *Notoryctes* spp.), while the American opossums show that it is possible for marsupials to survive and thrive alongside a varied population of eutherians.

2: Development and Morphology

Differentiation of male and female genital tracts

The morphogenesis of the accessory reproductive organs is described in standard textbooks of comparative anatomy, and need not be reviewed in detail here. A brief description of the differentiation of the male genital tract by development of the Wolffian (mesonephric) duct system, and that of the female tract from the Mullerian (paramesonephric) ducts, in relation to vertebrate phylogeny, is to be found in Brambell's book on the development of sex in vertebrates.[38] The role of gonadal hormones in normal sexual differentiation, however, has become much clearer in recent years, especially with the introduction of certain cytogenetic techniques and remarkable advances in organ culture—that is, the growth of whole organs *in vitro*.

In brief, the situation is, approximately, that the genetic sex of the individual determines whether the primordial gonad develops into a testis or an ovary. In mammals, the presence of a 'Y' chromosome leads to testis development, and in the absence of a 'Y' chromosome the indifferent gonad becomes an ovary. During a critical period, long before puberty and often during foetal life, the testis secretes androgen. As far as is known at present, the newly differentiated ovary secretes no hormone, and the available evidence suggests that the mammalian embryo of either genetic sex has the potential to develop male characteristics if androgen reaches it during the critical stage of development.

The significance of this precocious steroid secretion is not confined to its influence on the morphogenesis of the reproductive tract, for it is also responsible for 'imprinting' on the developing brain the pattern of activity that determines the differentiation of male or female pituitary function and, also, that of male or female sexual behaviour. These latter aspects are dealt with later (p. 62).

The sequence of events in normal sexual differentiation has been followed in organ culture preparations by Dorothy Price and her colleagues. They have succeeded in dissecting out the genital ridge and gonads of guinea-pig foetuses, together with the covering sheet of peritoneum, and growing the whole complex *in vitro* for nine days or longer. The guinea-pig was chosen because of its long gestation period (about 65 days) and the relatively long period (about 9 days) occupied by the changes under study. The sheet of peritoneum is, of course, transparently thin. When it is extended the gonads and genital tract are held in it and retain very nearly their normal relations to one another. In the technique developed by Professor Price, the tissue is manoeuvred on to lens paper, to which it adheres sufficiently to remain extended, and it can be examined or photographed by transmitted or incident light. In transmitted light, the duct systems show clearly.[306] The delicacy of the technique is indicated by the fact that the excised tracts are only 3 to 5 mm in overall length.

The critical phase of genital duct differentiation in the guinea-pig is from about 30–38 days p.c. The testis is recognizable, and testicular cords are formed, at about 24 days. When the testis was excised and removed at 26 days, the Wolffian duct failed to develop further, but from day 29 onwards removal of the testis did not prevent the Wolffian duct system from continuing to develop. It follows that this 3-day period must be the critical time in determining the development of the Wolffian duct system in the guinea-pig. When the testis was excised (in the organ culture preparation) and immediately replaced, the Wolffian duct developed normally.

A recognizable Fallopian tube is formed from the Mullerian duct in the female foetus from about 31 days p.c., in organ culture or in the intact foetus. Removal of the ovary has virtually no effect on this process. The question therefore arises: will testicular androgen inhibit Mullerian and promote Wolffian duct growth? When the ovary was removed, and a testis put in its place, Mullerian duct development continued normally, although Wolffian duct development took place alongside it if the testis implant was made early enough. These results, and others, suggest that foetal testicular androgen is responsible for initiating the development of the Wolffian duct system beyond the indeterminate, or ambi-sexual stage. They also suggest that early development of the female genital tract is 'anhormonal'. The observation that the presence of testis tissue stimulates the Wolffian duct but does not inhibit Mullerian duct development[306] confirms the results reported by Dantchakoff 30 years earlier.

In her experiments, Dantchakoff implanted or injected testosterone propionate into the amniotic or peritoneal cavity of female guinea-pigs 20 days after mating. Her results, and those of Price, appear to conflict, at first sight, with the results reported by Jost and others[210] which provide evidence that whereas androgen administration does not suppress Mullerian development, implants of embryonic testis tissue will do so. The implication of these experiments is that the testis produces androgen to stimulate Wolffian development and also produces something else to inhibit Mullerian development. The fact that testis transplants did not have this effect in Price's experiments perhaps implies that the inhibitory effect is exerted at a different time. Alternatively, it may be that this effect can only be exerted in the intact organism, or at least in the presence of organs not included in the organ culture preparation.

The work of Price and her collaborators stems in part from that of Lillie, to whom the hormonal theory of sex differentiation of mammals is chiefly due. Lillie[223] drew attention to the freemartin condition in cattle, and suggested that the intersexuality of the freemartin (a heifer born co-twin to a bull) was due to hormones from the male foetus reaching the female one because of vascular anastomoses on the allanto-chorionic surface. Recent work has shown that a number of cells are transferred or exchanged between such synchorial twins, presumably by way of the allantochorion. Whether this is so or not, and whether, if it is so, the resulting chimaerism explains the intersexual condition of the freemartin, has not yet been resolved.[292] The modern cytogenetic techniques that make possible the detection of such chimaeras have of course been applied to other species as well as cattle, and have been particularly valuable in the study of human intersexes.

Intersexes of various species are being widely studied at the present time, partly because of their clinical importance, but also in the hope that such investigations will help to explain the course of events in normal sexual differentiation; this, indeed, was the chief motive in Lillie's work on the freemartin. Naturally occurring intersexes in the Saanen breed of goats, genetically associated with hornlessness, have been investigated.[157] All the intersex individuals were found to be genetic females, the phenotype varying from almost normal female to almost normal male. The gonads usually resembled testes, and secreted testosterone, but no germ cells survived in them. The animals behaved like males, and the seminal vesicles contained fructose and citric acid. The condition was attributed to the retention of the medullary (testis-forming) sex-cords in the developing gonad. It was postulated that the gene for the polled (hornless) condition, or some genetic material always carried with this gene, acts like a 'Y' chromosome which, it had already been suggested, is not itself a direct male determinant, but functions by 'switching on' a male-inducing factor that is carried on the 'X' chromosome.

The suggestion that the foetal testis produces two 'inductor' substances, one of them androgen, has been put forward again very convincingly in work based on experiments involving drug-induced intersexuality in rats, dogs and rabbits.[273] The second 'sexual inductor', whose existence is postulated, is referred to as 'factor X'. Experiments involving castration or androgen administration can be used to compare the normal condition (presence of both factors) with that which results from the absence of both and with that which results from the presence of androgen when 'X' is absent (as in the work of Jost and others, already referred to). By using an 'anti-androgen' (a drug which inhibits the effects of steroidal androgen; see also p. 66) they were able to study the effects of 'X' in the (effective) absence of androgen. When this was done, both the Wolffian and Mullerian ducts of male foetuses regressed —the former presumably through lack of androgen, the latter from lack of factor 'X'.

The sharply defined critical periods (or 'determinant times') illustrate the 'rigid schedule'[273] of embryonic development. The great variability of form in intersexes can be accounted for by this fact. For the effect of a 'fault' in development will vary according to the stage of development at which it occurs. The organs whose development has already been determined will not be affected, while those that are still undifferentiated may be inhibited or rendered abnormal. An understanding of the factors involved, by no means complete as yet, is of great clinical importance, not only in respect of abnormalities that arise spontaneously but, perhaps even more vitally, in respect of possible untoward side effects of synthetic drugs. The screening of such drugs on behalf of the manufacturer, in a wide variety of experimental animals, has already become a considerable industry.

Early experiments in hormonal sex reversal showed that relatively large doses of oestrogen administered to a pregnant mammal may result in the partial feminization of the genitalia of male foetuses. At first sight, this well established result appears to conflict with what has been said about the passivity of the ovary in sex differentiation, and the absence of foetal ovarian oestrogen in contrast to the presence of foetal testicular androgen. It is thought, however, that the effect of oestrogen reaching the foetus via the placenta is due, not to its having a direct effect on the foetal accessory organs, but to its inhibiting the secretory function of the foetal testis.

The role of the primordial germ cells in sex differentiation is still not absolutely clear. Most probably, however, they play no part in the early differentiation of the genital tracts. The gonadal tissue apparently differentiates according to its own genetic constitution, and medullary (male) or cortical (female) dominance is established before the primordial germ cells reach the gonad. When they do so, they perhaps 'home' in the medulla or cortex according to the sex. The gonad remains sterile if the primordial germ cells do not reach it, but differentiation of the male or female genital tract proceeds normally.

Ovarian development and oocyte formation

It is not within the scope of this book to describe the embryonic development of the ovary in detail, but it is necessary to have a clear picture of the structure of the organ and its components. The development of the vertebrate ovary may be divided into four main phases:[139] the first consists of the migration of the primordial germ-cells from their site of origin, which is always extra-gonadal, to the genital ridges in the dorsal wall of the coelom. The second phase consists in the formation of gonadal primordia which are anatomically similar in both sexes (the indifferent gonad), and during the third phase a peripheral cortex and central medulla are differentiated within the gonad. Differentiation of the sexes marks the fourth phase of development; it involves the elaboration of the cortex and the involution of the medulla in the female, whereas the converse is true of the male. The internal genitalia of the foetus can usually be distinguished anatomically as male or female in the latter third of pregnancy.

The primordial germ cells are usually identifiable as such during their migration, but soon after reaching the genital ridges they lose their distinguishing cytological features, and their subsequent fate was for long a matter of argument. The controversy centred on the problem whether or not the oocytes that were derived directly from the primordial oocytes were supplemented, or replaced, by others formed in adult life. The argument arose from the fact that oocyte formation involving the characteristic cytological changes of the prophase of the heterotypic (reduction) division has not been observed after puberty in any species of mammal, with the notable exception of some lemurs. To some, it seemed incredible that a long-lived animal should depend, throughout its reproductive life, on the stock of oocytes laid in about the time of birth. To others, it was inconceivable that in mammals alone, and in the female sex only, germ cells should be formed without revealing any cytological evidence of meiosis.[41] It is now generally accepted that the primordial germ cells not only give rise to all the definitive gonocytes in both sexes, but also play an important part in the organization of the gonads. Their locomotion by amoeboid movement has been described from living material, and recorded by cinematography.[31] The literature concerning their movement and multiplication in mammals and other vertebrates, during the first of the four phases of ovarian development already described, has been reviewed.[167]

During the second phase of ovarian development the primordial germ cells congregate in the middle part of the genital ridge of each side; this portion of the ridge thickens and forms the gonad. In the female, the cranial portion of the genital ridge forms the suspensory (ovarian) ligament, and the caudal portion forms the utero-ovarian ligament. The connection between the genital ridge and the meso-nephros forms the mesovarium, through which nerves and vascular elements reach the ovary.

Ovarian innervation is autonomic; it is supplied from the renal plexus and the superior and inferior hypogastric plexus, together with fibres which probably take origin elsewhere. The ovarian and uterine nerves run together and have many inter-connections. The ovary itself is abundantly supplied with nerve fibres but they appear to be related only to the blood vessels and fibromuscular tissues. The follicles and secretory tissues are said to be devoid of a nerve supply. Ovulation has been induced hormonally in a completely denervated ovary, the smooth musculature of the whole Mullerian tract can perform rhythmic con-tractions in the absence of nerve impulses and parturition has been recorded after section of all nerves to the uterus. The nervous system has therefore been thought to play a minimal part in regulating ovarian or uterine activity. However, the cyclic changes in blood supply must involve the accompanying nerve fibres and the utero-ovarian innerva-tion may well be found to have a more important role in the control of ovarian periodicity than has hitherto been recognised.

The similarity between the terms 'mesonephros' and 'mesovarium' is fortuitous, and may be confusing. The mesonephros is the 'middle kidney' coming, as it were, between the pronephros and the metanephros. 'Mesovarium' is cognate with 'mesentery', 'mesometrium', etc., and signifies not the organ itself but its suspensory ligament—the fold of peritoneum in which it is suspended in the coelomic (peritoneal or abdominal) cavity. The fold of peritoneum of course encloses a thin sheet of connective tissue through which run the ovarian artery and vein, a number of lymph channels and the ovarian nerves.

Whether the indifferent gonad will develop into an ovary or a testis —that is, whether the cortical or the medullary elements will develop— is primarily determined by the genetic sex of the individual, but it is subject to modification by a variety of internal and external factors, and it can be altered by experimental interference. The mode of action of the sex chromosomes, and their interaction with the autosomes and with the cytoplasm in bringing about the differentiation of the gonad, remains unknown.

During the early stages of differentiation of the ovary, the oogonia derived from the primordial germ cells multiply rapidly by mitosis. The cells of the 'germinal epithelium' (which is simply the epithelium covering the surface of the ovary and continuous with the peritoneum) also multiply, and according to the classical descriptions, ingrowths from this epithelium (Pflüger's tubes) provide the elements from which the 'primordial follicles' are formed. This term is used to denote the simple capsule, one cell in thickness, which forms itself around each primary oocyte about the time it enters upon the long resting stage of the prophase of the first reduction division (meiosis). Thus the history of the germ cells within the ovary may be summarized as follows: the primordial germ cells divide mitotically an indefinite number of times, giving rise to a population of oogonia. Each oogonium is transformed into a primary oocyte, a cell which becomes enclosed in a follicle and is destined either to degenerate or to complete the reduction division and eventually to provide an ovum. The stages in this process are shown diagrammatically in Fig. 1, and Plate I shows a photomicrograph of primordial follicles in a guinea-pig ovary. Plate II is an electron micrograph of one such follicle. It is at this stage in the development of the oocyte, in the prophase of the first reduction division, that its nuclear assortment is halted and, as we have seen, it is about this time that it acquires a capsule of epithelial cells which constitute its follicle. In the majority of mammals this stage is reached—that is, all the surviving oogonia have ceased to multiply and have entered upon the prophase of the reduction division and may therefore be called primary oocytes—either before or soon after birth. From this time until shortly before it is released from the ovary, the oocyte undergoes little change other than an increase in its cytoplasm. The nucleus remains large, and its chromatin pattern is

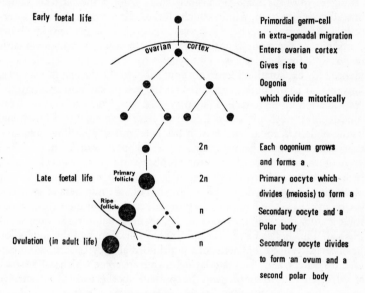

FIG. 1. Mammalian oogenesis.

characteristic; it is sometimes referred to as the 'germinal vesicle'. The follicle surrounding the oocyte will grow and will undergo great changes; the oocyte is insulated from change within the follicle, as will be described later.

Intercellular bridges between oocytes

When the primary oocytes are first formed, they lie in clusters within meshes of stromal tissue. Primordial follicles are formed as individual oocytes are enveloped by the stromal cells, which thus isolate the oocyte and completely enclose it within a capsule. While they are still in clusters, so that the surface of one oocyte is likely to be in contact with others, the cytoplasm of adjacent oocytes is often found to be continuous across 'intercellular bridges' as though the last few mitotic divisions had been incomplete. Such connections (as distinct from desmosomes, which have sometimes been described as 'bridges') occur in few tissues, but are characteristic of germ cells in the male. The morphological arrangement of spermatogonia is extremely orderly, the germ cells lying in well defined arrays, within which all the cells are connected by inter-cellular bridges. It has been suggested that this arrangement, involving continuity of the cytoplasm throughout the group of spermatogonia, is responsible for the high degree of synchrony observed in their develop-ment. This explanation is scarcely applicable to the developing oocytes, however. It is, of course, arguable that the cytoplasmic bridges have no real function in the female. They may exist simply as a carry-over from the inherent mechanism which is more fully developed and functional in the male. On the other hand, it has been rather plausibly suggested that the ovarian arrangement may serve a real but very different function. Although some degree of synchrony is observed in oocyte development, it is not comparable with that seen in spermatogenesis, and whereas events in the testis tend to the production of vast numbers of gametes, the germ cells of the female undergo a reduction in numbers throughout their history, beginning as soon as the last oogonial mitoses are accom-plished. This in itself suggests a process of selection, and the suggestion has been advanced that some of the primordial oocytes serve as nurse-cells for others, and that material is drawn from some which will at some stage degenerate, to supplement the contents of others which will survive at least for a time.[328] This hypothesis is attractive in that it ascribes a function to an observed structure, but one must bear in mind that the actual transfer of material from the one cell to the other has not been demonstrated.

Oogenesis in adult lemuroids

The multiplication of oogonia, and the characteristic processes of oogenesis, are continued into adult life only in a few of the lower pri-mates. The phenomenon has been observed in several species of lemur and loris, but it is not yet known whether it occurs in all of them, or whether it occurs in the tarsoids (see p. 75). In reviewing the work of earlier authors Brambell referred to the problem of the origin of these oogonia and remarked that further investigations would be required in order to determine 'Whether the new germ cells arise by the transformation of epithelial cells, as was concluded by Gérard and by Rao and as seems probable, or from oogonia derived from primordial germ cells which have persisted as such into adult life, as is possible at least theoretically'.[41] Some 'further investigation' has been done; and the investigators incline to the second alternative outlined by Brambell in the passage quoted above. They confirm that oogenesis is not interrupted by pregnancy, but they show that it fluctuates with the oestrous cycle. The number of oogonia in the ovaries varies with the oestrous cycle, suggesting that oogenesis is under endocrine control.[11] It is affected by ovarian rather than hypophyseal hormones. These observations may have no significance with regard to pre-natal or post-natal oogenesis in other species, for oestrogen is decidedly mitogenic—that is, it stimulates cell division—so the fact that oogenesis, a form of cell division, is stimulated by it in adult life, does not necessarily indicate that this hormone plays a part in stimulating oogenesis at a much earlier stage of development. In the lemurs which illustrate oogenesis in adult life, the oogonia lie in 'nests' of epithelial cells in the outer zone of the ovary. These are ingrowths from the 'germinal epithelium', corresponding to Pflüger's tubes in the early ovarian development described above. Ioannou[206] 'occasionally' found oogonia in the germinal epithelium, but in his opinion 'there is no reason to believe that such cells are derived from transformed epithelial cells'. Anand Kumar observed that, in two species of *Loris*, the cells of the germinal epithelium were seen to divide infrequently.[11] He agreed with Ioannou that the primary oocytes formed in adult prosimians arise from oogonia produced by mitotic division of primordial germ cells which persist from foetal stages, but he has emphasised that the persistence of oogonia, and their apparent conversion into oocytes, does not conclusively prove that these oocytes constitute the female gametes that are actually fertilized.

Oocyte numbers

Once it had become generally accepted that oogenesis (or oocytogenesis to be more precise) ceases within a short time after birth in the generality of mammals, the process was re-examined in a number of species. Oocyte numbers have been estimated in some of the laboratory species, in the human and in the rhesus monkey among the primates, and in recent years some of the larger domestic animals have been studied.

One of the first species to be studied from this point of view was the rabbit, and its quantitative re-examination has led to a curious anomaly in the literature. It was among the species included in the classical work by de Winiwarter of Liège, which is widely quoted as showing that the definitive germ cells (oocytes) are present in the ovaries by or about the time of birth.[391] A recent paper in which no reference to de Winiwarter's work is made, shows that oogenesis continues during the first two weeks after birth in this species.[295] One subsequently finds this paper referred to as the description of an exceptional case—'The first stages of meiosis are completed prenatally in all species reported with the exception of the rabbit'.[29]

This is a good example of the kind of apparent contradiction that often arises when the same phenomenon is seen from different viewpoints. De Winiwarter in fact stated that oogenesis continued until ten days after birth, but emphasis is inevitably laid upon the unexpected, and at that time the surprising thing was that oogenesis should cease so *early* in life. Sixty-five years later, when this had long been established, and oogenesis had been shown to be completed before birth in a number of species, the surprising thing was that it should persist so long *after* birth in the rabbit. So, what appeared precocious to the earlier investigators later came to seem tardy.

In view of the incredulity that the 'early stock-pile' of oocytes aroused, it is pleasant to read with what conviction and clarity de Winiwarter gave his interpretation. Having said that the oocytes reach the resting (diplotene) stage 'jusqu'à environ 10 jours après la naissance', he goes on to say, 'A partir de ce moment, il ne se forme plus de nouveaux ovules, du moins chez le jeune animal. Si plus tard, chez l'animal adulte, il se produit une néoformation d'oeufs, ce fait ne sera démontré d'une manière décisive que pour autant que les prétendus ovules de nouvelle formation montreront les métamorphoses nucléaires caractéristiques des premiers stades du développement de l'ovaire . . . Je suis convaincu, pour ma part, que cette recherche donnera un résultat négatif'.

More recent work has shown that oogenesis is completed during foetal life in the cow,[121] but continues into neonatal life in the pig.[29] Among the species so far studied in this way (excluding the lemurs of course) it is clear that oogenesis takes place, and is completed, during early stages in the development of the individual. Development proceeds further in the uterus in some species than in others, and those born in a relatively undeveloped state (rabbit, pig, rat) tend not to have completed oocyte formation by the time they are born, whereas those that are carried longer in the uterus in relation to their development (cow, guinea-pig) tend to have done so.

The actual number of germ cells present in the ovary, at successive stages of development, has been estimated for several species including human, rhesus monkey, guinea-pig, rat, mouse, cow and pig. The number rises, of course, as the oogonia undergo mitosis, the increase is slowed as some enter meiosis, and further slowed as some oocytes degenerate. When mitotic activity ceases the total can no longer increase, and further 'atresia' (p. 26) leads to a continuing decline in the total number. Later, some of them mature and are ovulated at regular intervals, but this contributes relatively little to the gradual attrition of the original stock of oocytes.

In the human, there is great individual variation; typically the total population of germ cells reaches a peak of nearly seven million in the fifth month of pregnancy. It then declines until there are some two million at birth, and half of these are atretic. Of the one million normal ones present in the newborn, about 300,000 remain at the age of 7 years. The corresponding figures for the rhesus monkey follow a very similar pattern of change, but there are about half as many germ cells or oocytes at each stage.

In the rat, oogonia are found in mitosis up to $17\frac{1}{2}$ days p.c. (gestation 21 days) and the total number of germ cells reaches about 71,000 by that time. From then onwards, the number declines; diplotene oocytes are found about 2 days after birth, when the total germ-cell number is down to about 19,000. In the guinea-pig, a peak total of about 105,000 is reached at 41 days p.c. (gestation 66 days) and the 12-month-old animal has some 13,000 oocytes. Of these, however, the great majority have a peculiar type of contracted nucleus, and only about 1,600 have not undergone this change by this time. It is not known whether this nuclear contraction is a form of degeneration, or whether the oocytes with this characteristic, or the uncontracted ones, are viable.

A comparison of the rat with the guinea-pig, or the human with the rhesus monkey, might suggest that the peak total number of germ cells is related to body size, but the cow has fewer than the primates. The total was estimated to reach a peak of 2,700,000 at 110 days p.c. in the cow (gestation 270 days). Some days before term, the foetal ovaries contained relatively few oocytes—between 50,000 and 90,000. In the pig, the total reaches a peak of over a million germ cells at 50 days p.c., and there are some 500,000 in the ovaries at the time of birth.

The questions naturally arise: why are oocytes produced in such prodigious numbers? Why are so many squandered in atresia? Is reproductive capability terminated by exhaustion of the oocyte stock? Atresia is discussed elsewhere, in relation to follicular growth and regression; the first two questions are perhaps answered (though rather unsatisfactorily) by the suggestion that it is not the oocytes that are required in such numbers, but the follicles, whose formation they induce.

As to the possibility that the reproductive span ends when the stock of oocytes is exhausted, one can say that this is certainly not the only factor and probably not the chief one. Experiments have been performed in which the number of oocytes was reduced by some means in animals where subsequent reproductive performance was compared with normal controls. The earliest experiment of this kind was performed by John Hunter, who published an account of it in 1787.[202] This account is often quoted, not always accurately. It describes the farrowing record of two sows, one of which was semi-spayed (one ovary removed). The number of piglets in each litter was similar in each sow until, after the eighth litter, the semi-spayed one ceased to breed, having produced 76 piglets in all. The intact sow produced 87 piglets in her first eight litters, but went on to bear a further five litters, comprising a further 75 piglets. Her 13 litters totalled 162 piglets, an average of 12·5, and the twelfth and thirteenth pregnancies produced 16 and 19 piglets respectively. Thus for eight pregnancies the sow with only one ovary almost kept pace with the sow that had two, but she only produced about half as many offspring over the whole reproductive span. Hunter therefore reached the impeccable conclusion that although 'the constitution at large has no power of giving to one ovarium the power of propagating the equal to two', it did influence ovarian activity in such a way 'as to make it (the single ovary) produce its numbers in a less time than would probably have been the case if both ovaria had been preserved'.

In more recent times it has been clearly demonstrated that unilaterally spayed animals produce litters of approximately normal size. The implication is that the gonadotrophic stimulus operates entirely on the single ovary and that the normal complement of follicles ovulates whether distributed between two ovaries or concentrated in one. It would appear, however, that Hunter's experiment, covering the whole reproductive span, was not repeated in any species until about ten years ago, when Jones and Krohn recorded the effects of unilateral ovariectomy on the reproductive lifespan of twenty mice, whose performance was compared with that of intact litter-mate controls.[209] As in Hunter's experiment, the reproductive lifespan was curtailed and the total number of offspring born was approximately halved. But was this because the semi-spayed mice ran out of oocytes? Apparently not, for the number of oocytes left in the ovary, some weeks after the last litter was born, was similar to the number found in each ovary of intact mice of the same age.

The anatomy of the uterine cervix in the mouse is such that embryos from one uterine horn cannot migrate through it to the other horn. Because of this, ova shed from either ovary must develop in the uterine horn of the same side. When one ovary is made to ovulate twice its normal output of ova, as in the unilaterally spayed mouse, the uterus is necessarily crowded, and this 'overloading' may account for the premature ageing. It would therefore seem logical to repeat this type of experiment with a species in which embryos can move freely between the uterine horns before becoming implanted, but the anatomical barrier to inter-cornual distribution of embryos that exists in the mouse also applies to the rat, rabbit and guinea-pig. As the nature of the experiment calls for the use of a polytocous species, the only common one that fulfils the requirements is the pig. It has not been thought worthwhile to do this experiment in modern times, with significant numbers of animals, because the time and labour involved would be disproportionate to the probable value of the information gained. For it is difficult to believe that oocyte exhaustion is responsible for senile sterility, or for menopause in man. On the other hand, it would certainly be interesting to discover whether John Hunter's single pair of sows really represented a difference between hemi-spayed and intact pigs, or whether he happened to select one animal whose 'constitution' was such that she would have produced only 8 litters, comprising 76 piglets, even with her original complement of 'ovaria'.

Growth of the follicle and oocyte

Growth of the oocyte bears almost no relation to adult body size, whereas the volume of the mature follicle, which is mainly determined by the size of the antrum is, in general, related to the adult body size of the species.

The epithelial cells forming the primordial follicle soon begin to multiply while retaining their orderly arrangement as a wall around the oocyte. When this wall is some six cells thick, the immediately surrounding stromal cells become differentiated to form a case (theca) around them, and henceforth the theca is regarded as an integral part of the follicle. Its development is apparently induced by the presence of the inner epithelial cells, and is perhaps brought about by a secretion from these cells, which constitute the membrana granulosa.[110]

The differentiation of theca and granulosa is of great significance in ovarian function; in all probability the thecal cells are the main source of oestrogen and the granulosa cells the main source of progesterone. The theca and granulosa are sharply demarcated from each other by another 'membrane' known as the membrana propria, a term that implies its origin as an epithelial basement membrane. It was, in fact, long regarded as such, but it has since been shown to be composed of the true basement membrane backed by a collagenous layer of connective tissue. The membrana propria appears to prevent the penetration of vascular elements from the theca into the granulosa until after ovulation, when blood and lymph vessels do penetrate it and the granulosa is transformed by 'luteinization' (see p. 49). The theca of the growing and mature follicle is extremely vascular, whereas the granulosa is completely avascular. The oocyte enters upon its final maturation changes as soon as it is freed from the follicle, and it may well be that the membrana propria plays an essential part in staying this process and in controlling the timing of events at ovulation. It has received relatively little attention, and both its origin, and the factors that cause its sudden dissolution, are somewhat obscure.

After the separation of the theca and granulosa the cells of the latter multiply only very slowly, if at all. The theca continues to grow, however, and the follicle, which remains spherical, swells. A rift appears within the granulosa and forms, at first, a fluid-filled slit. As the follicle expands, this rift, called the antrum, is enlarged. In the mature follicle of most species it accounts for most of the volume occupied by the follicle. The oocyte remains enclosed within a clump of granulosa cells and this clump (the 'discus proligerus') remains attached to the granulosa cells that line the follicle wall (Plate III). When the follicle ruptures and the oocyte is shed at ovulation, most of the cells of the discus go with it and form the 'cumulus oophorus'. They are shed during the oocyte's passage down the fallopian tube. The innermost layer, the cells of which are intimately connected with the zona pellucida, forms the 'corona radiata'. This is an apt descriptive term, for the nuclei appear to move away from the oocyte, so that the cells are attentuated as they remain, for a time, attached to the zona pellucida (see below), probably by the cytoplasmic processes that formerly penetrated it. The mature follicle of the mammal was first described by a Dutch physician, Regnier de Graaf in the seventeenth century; hence the vesicular follicle (one with an antrum) is referred to as a 'graafian' follicle.

Two further terms must be introduced here. The growing follicle compresses the stromal cells immediately surrounding it, forming a somewhat ill-defined 'theca externa'. This layer is often described as an integral part of the follicle, the wall of which is said to consist of a 'theca interna' and 'theca externa', but its appearance seems to me to indicate that it is formed simply as a result of compression by the swelling follicle. The word 'theca' used alone may be taken to refer to the theca interna. The cells of this layer always have an important endocrine function, and the tissue is sometimes so obviously differentiated from the surrounding cells that it is described as a 'thecal gland' (p. 29).

The zona pellucida

The oocyte reaches its maximum size while the follicle is still very small —before the appearance of the antrum. About this time a non-cellular layer, the 'zona pellucida', begins to form around the oocyte. The cell wall, or limiting membrane, of the oocyte is the vitelline membrane, and the zona pellucida appears between it and the innermost granulosa cells. Its origin is somewhat obscure but its mucopolysaccharide constitution and its ultrastructure have been extensively studied. It thickens as the follicle grows, and it is traversed by radial canaliculi which extend into it from both inner and outer surfaces. Processes from the oocyte and from the granulosa cells extend into the thickness of the zona within the canaliculi, and possibly meet within it. Processes from the innermost follicle cells of living human eggs appear to pass completely through the zona and allow material to be conveyed into the perivitelline space.[348] These structures have been intensively studied in recent years with the aid of the electron microscope (see p. 35), but their nature and function have not yet been completely elucidated. Indeed, it is not clear whether the zona has any function at this stage, but it certainly forms a robust protective layer around the egg when the latter is shed. It remains in place after ovulation, and has to be penetrated and traversed by the spermatozoon that fertilizes the oocyte. It has been conjectured that the spermatozoa enter the zona by the canaliculi but this has not been substantiated. The egg is not only fertilized within the zona, but begins to develop and grow within it. The zona is stretched and becomes thinner as the 'egg' enlarges; it is retained until the blastocyst stage of development is reached, and then it is shed, either by dehiscence or, in many species, by a more gradual dissolution.

To return to the follicle: as it grows it sinks into the stroma of the ovary, away from the surface. The relatively few follicles that are destined to ovulate make, as it were, a return journey, for during the later stages of their development they move out again towards the surface, and usually protrude slightly before they rupture. In the sow, where many follicles ovulate in each ovary, one has the impression that some of those that mature but do not ovulate, fail to do so simply because there is not enough room for them at the surface of the ovary.

A residual stock of primordial follicles remains near the ovarian surface, and they form a very conspicuous layer in the ovaries of some species, such as that of the cat as shown in Plate III.

Ovulation and atresia

Ovulation

The ovarian changes that culminate in ovulation consist in the growth of one or several graafian follicles under the influence of gonadotrophin released into the bloodstream by the anterior pituitary gland (also referred to as the adenohypophysis or pars distalis). These follicles are selected, as it were, from among those that have reached a relatively advanced stage of growth and possess an antrum. The means of selection is not clear, but augmentation of the amount of available gonadotrophin (follicle-stimulating hormone, FSH) results in the maturation of a larger number. Similarly, if one ovary is removed, the number of maturing follicles is not correspondingly halved; more are stimulated in the remaining ovary than would normally mature, and as many ova are shed as in normal control animals.[241] This was the subject of John Hunter's famous experiment with sows which was described (p. 20) in connection with the question of the stock of oocytes available during the reproductive span. In these circumstances the remaining ovary increases in size, a phenomenon long recognized as 'compensatory hypertrophy'. The fact that the number of follicles that ovulate at one time is characteristic of the species was stated as the 'Law of Follicular Constancy' by Lipschütz in 1928.[226] The most highly purified preparations of either follicle-stimulating hormone (FSH) or luteinizing hormone (LH) have been shown to be ineffective when administered alone[229] so that successive stages of the ovarian cycle seem to be initiated by changes in the ratio of these hormones in the circulating blood, rather than by the total replacement of one by the other. FSH is predominant during the phase of follicular growth, and LH is generally regarded as inducing ovulation and corpus luteum formation (see p. 49). The process of mammalian ovulation has not been completely elucidated, either in terms of endocrinology or with respect to the mechanics involved. It is clear that the follicle does not burst under intra-follicular pressure like an over-inflated balloon; indeed the follicles of pigs, cows and sheep lose their turgor before they actually rupture, and this may well be true of other species. The follicle wall is said to include smooth muscle fibres, but substances that would be expected to cause such fibres to contract, such as oxytocin, do not cause or hasten ovulation. On the other hand, it is evident that the follicle must have reached a certain state of development before it can respond to the ovulating hormone. It attains a responsive state some time before the ovulatory signal is received, and it retains this capacity for a period, but eventually loses it. Descriptions of the gross morphological changes in the follicle wall before and during ovulation, were reviewed by Brambell in 1956[41] and, as far as I am aware, little has been added to our knowledge of them since then. The electron microscope has not been applied to the problem on any large scale, but it has been used to study changes in the rupture point of the rabbit follicle during ovulation. The wall becomes markedly attenuated over the apex of the follicle, and the granulosa layer disappears entirely from this area. The cells of the thecal layers and the overlying germinal epithelium become progressively disassociated, and collagen fibres become sparser, as if the cellular and fibrous matrix were being dispersed, perhaps under the action of hydrolytic enzymes. Eventually the wall becomes so thin that the slight intra-follicular pressure is sufficient to expel the egg and the cumulus surrounding it. The collagen is mainly concentrated in the outermost layer underlying the 'germinal epithelium' —i.e., in the tunica albuginea and not in the follicle wall proper, and

B

this layer appears to constitute the principal restraint as follicular rupture becomes imminent. 'Even though most pre-rupture follicles were taken within minutes of estimated rupture, in no instance was the apex completely devoid of collagenous tissue from the thecae and tunica. Nor had the germinal epithelium sloughed off from some of the most distended follicles. On the other hand, the stratum (membrana) granulosum, membrana propria and adjacent thecal capillaries were sometimes absent from the apex of ovulating follicles. This observation suggests the apical translucent area develops when the broken granulosa and membrana propria retract to the edges of the protruding cone. Sometimes the ruptured capillaries at the surface of the membrana propria rescind with the granulosa cells. This condition is probably responsible for the frequent appearance of the cone (stigma) as a circumscribed avascular area'.[122] The dispersion of granulosa cells is characteristic of atresia and is clearly seen in 'over-ripe' follicles or in cystic follicles that fail to rupture. Pre-ovulatory changes in the theca are confined to 'ripe' follicles, and a rupture point appears to form over the area where the theca of such a follicle closely approaches the overlying tunica albuginea.

It would be interesting to examine in similar detail the process of ovulation in animals such as the tree shrews (Macroscelididae) and the tenrecs (Centetidae) in which the ripe follicle is 'everted' so that the corpus luteum protrudes from the surface of the ovary on a stalk or pedicle of stromal tissue. In both groups, luteinisation of the granulosa (see p. 29) starts before ovulation, and in the tenrecs the swelling granulosa occludes the antrum so that there is no cavity in the mature follicle.[172]

Atresia

Most mammalian oocytes are destined to degenerate, and their follicles cease to exist as such. The follicle cells do not necessarily, or even usually, degenerate and disappear, but apart from a very few that may grow, and even form normal-looking corpora lutea, the follicles of degenerating oocytes fail to mature and rupture. It would seem that the term 'atresia', as applied to the ovarian follicle, originally implied simply a failure to rupture. It is also possible that it was originally used to describe a failure to develop an antrum (although by derivation it clearly refers to an orifice and not a cavity) but its exact historical origin appears to be lost. Whatever its origin, however, this term is now often used to cover all the changes concomitant with the degeneration of an oocyte. Ingram, for example, used it to represent 'the process or processes whereby oocytes are lost from the ovary other than by ovulation'.[204] He emphasized its importance in normal ovarian physiology by pointing out that of the million or so oocytes in the human ovary at birth, only about 400 will be ovulated before menopause, when the stock is nearly exhausted (it should be noted, however, that reproductive activity has usually ceased before this, and is due to other causes).[208]

The cause of atresia is not known. Cytologically it probably begins with changes in the oocyte itself, but this is not certain. Nor is it yet possible to distinguish between those oocytes destined to be ovulated and those destined to degenerate. When oocytes are released from the ovary by dissection they enter upon the nuclear changes that normally precede fertilization, just as they do after their release by follicular rupture at normal ovulation or under the influence of gonadotrophin administered experimentally.[66, 116] These observations do not indicate that a large proportion of oocytes are capable of development. During atresia, the degenerative processes involve the granulosa cells as well as the oocyte. Thecal cells of atretic follicles, on the other hand, typically undergo hyperplasia and form, in many species, the well-defined and physiologically important interstitial tissue.[192]

Interstitial tissue

This rather nondescript appellation tends to encourage neglect of a most intriguing component of the mammalian ovary. It has been established, in recent years, that its cells have a well defined role in steroidogenesis. That they were secretory was deduced from their histological and cytological character by early histologists, notably by Bouin who, in 1902, wrote of 'the two glands of internal secretion in the ovary; the interstitial gland and the corpus luteum'.[34] Whereas the graafian follicles and corpora lutea are always conspicuous, interstitial tissue varies greatly in amount in different species. It is very prominent in the rabbit, and in many species of Insectivora, Chiroptera, Rodentia and Carnivora, but sparse in man and in Primates in general, as well as in Cetacea, Artiodactyla and Perissodactyla. It is very doubtful whether interstitial cells are entirely absent from the ovaries of any species. The tissue is very prominent in marsupials, and is identifiable in the ovaries of pouch young, before the formation of follicles.

Arguments concerning the origin of ovarian interstitial cells, as to whether they arose from the germinal epithelium, or from stromal or thecal elements, have been partially resolved by the realization that in some species at least, they may be formed from different sources at successive stages of development. In the rat, for example, it has been shown that 'primary' interstitial tissue is formed by ingrowing cords of cells derived from the germinal epithelium during early life, whereas near puberty and in adult life 'secondary' interstitial tissue is clearly formed from the theca interna of atretic follicles. Whatever the actual function of the primary type, it has been shown to be capable of producing oestrogen and androgen in response to hormonal stimulation.

In many species the interstitial cells enlarge in pregnancy; an extreme example is that of the water-shrew, where interstitial and luteal cells are indistinguishable.[307] Little attention appears to have been paid to the electron microscopy of interstitial tissue, but in the rat it has been shown that interstitial cells are separated from the endothelial cells of adjoining capillaries by 'sub-endothelial spaces' similar to those which characterize the corpus luteum. There seems little doubt that these cells secrete hormonal products into the bloodstream or into the lymphatic circulation. It seems probable that they are responsible for the hormone secretion that maintains the oestrous cycle in mice after all the follicular apparatus has been destroyed by X-rays.[280]

In the horse and the elephant, in which the foetal gonad hypertrophies in late pregnancy, apparently under the stimulus of maternal gonadotrophins (see p. 130), the increase in ovarian size is presumably in the 'primary' interstitial tissue. In the giraffe, where the foetal ovary becomes filled with corpora lutea, the situation is evidently different, probably because the follicles are further developed in this species by the time the stimulus reaches the gonads.

Formation and function of the corpus luteum

The corpus luteum, like many other mammalian structures, was described and named long before any function could be ascribed to it. The early history of the elucidation of its function has been summarized by Parkes and Deanesly.[285] In the latter years of the nineteenth century several investigators appear to have attributed a glandular function to the corpus luteum, and associated its presence with the suppression of ovulation. The idea of it as a gland of internal secretion necessary to the implantation and early development of the embryo, is attributed to Gustav Born, of Breslau who, in 1899, is said to have made the suggestion from his deathbed to his pupil Fraenkel. Conclusive proof of this theory was obtained by experiments with mated rabbits in which corpora lutea were ablated by cautery or eliminated by ovariectomy. Fraenkel published these findings in 1910,[138] by which time Bouin and Ancel[35, 36] had described the progestational proliferation of the uterine endometrium in the rabbit, and had demonstrated its dependence on the corpus luteum.

Follicular growth after the appearance of an antrum consists mainly in the enlargement of this fluid-filled 'space', but in the final stages before ovulation the cells of the theca interna hypertrophy and acquire a glandular appearance. The granulosa cells do not change in appearance until about the time of ovulation, but when the thecal blood vessels penetrate the 'membrana propria' and grow into the developing corpus luteum (as the ruptured follicle may immediately be called) the granulosa cells rapidly hypertrophy. As these changes are characteristic of the corpus luteum, the granulosa cells are said to 'luteinize' and the same expression has sometimes been used to describe the similar, but less marked, pre-ovulatory changes in the thecal cells. In some species in which the theca becomes very prominent it has been referred to as a 'thecal gland' after its hypertrophy. This expression is attributed to Mossman in describing the ovary of the pocket gopher, *Geomys bursarius*.[264] He showed that thecal hypertrophy coincided with the period of maximum output of oestrogen by the ovary in this species, and later suggested that the 'thecal gland' of the guinea-pig, also, was the principal source of oestrogen.[367] This was in line with Corner's conclusion that 'the most probable site of production (of oestrogen) in the ovary is the theca interna of follicles of all sizes'.[80]

With the development of methods for the chemical determination of small amounts of various steroid hormones, it became possible to identify sites of production with more precision. An attractive hypothesis put forward by Short was based on differences between the corpus luteum and the follicular fluid of the mare with respect to the relative amounts of different steroids they contained.[351] He suggested the presence of two cell types in the ovary. In his hypothesis both cell types are regarded as being capable of converting cholesterol to pregnenolone and thence to progesterone, but whereas one type (thecal) continues this process by 17-hydroxylation to form 17α-hydroxyprogesterone, and thence eventually produces oestrone and oestradiol, the other type (granulosal) cannot carry the process beyond progesterone except by reducing it to 20α-hydroxypregn-4-en-3-one, which was not found in follicular fluid.

Short's results, and conclusions, support those of Falck who, in a series of experiments with rats, transplanted theca interna cells, or granulosa cells, or both, into the anterior chamber of the eye, together with some vaginal epithelial tissue to serve as an indicator of oestrogenic activity.[128] Granulosa cells grew into 'luteal' cells but, in the absence of thecal cells, failed to cornify the vaginal epithelium; thecal cells caused cornification of the vaginal epithelium, but only if some granulosa cells were also present. There is therefore considerable support for the long-standing suggestion that oestrogen is produced by thecal cells and pro-gesterone by 'luteinized' granulosa cells. The ovary, however, is an extremely plastic organ. The thecal tissue of atretic follicles appears to be transformed into 'interstitial' tissue, and interstitial tissue may luteinize and become at least histologically indistinguishable from 'true' luteal tissues, as, for example, in the pregnant water-shrew[307] and in the stromal tissue of the foetal gonad (before it has been augmented by the atresia of vesicular follicles) of the horse[73] and of the elephant.[291] Furthermore, it is not certain, in many species, whether the 'luteal' cells of the corpus luteum are derived exclusively from the granulosa, or what proportion of them are derived from thecal cells. It seems probable that both granulosa and thecal cells contribute to the functional corpus luteum in most species, but it is possible that it is formed from granulosa cells alone in some species. I do not know of any well-founded descrip-tion of a species in which the luteal cells appear to be exclusively derived from thecal cells, but the possibility has been suggested in a study of a few specimens of Hyrax.[277]

There is little doubt as to the dual origin of the luteal cells in the cow, or in women. In the latter, it has been claimed that the cells of granulosal and thecal origin can be distinguished histologically in the earlier stages of pregnancy. In the cow, division and enlargement of cells after the first four days is confined to those which are in or near the ingrowing trabeculae of thecal tissue.[105] The centrally situated cells of granulosal origin enlarge immediately after ovulation and 'hold the fort' until the thecal cells have multiplied sufficiently to provide an adequate number of luteal cells. These, by their enlargement, account for most of the weight increase in the corpus luteum. This interpretation was based on a plentiful supply of exceptionally well preserved material; it is at variance with some earlier accounts in which the main luteal elements were seen as being granulosal, while cells of thecal origin remained smaller.

Among the laboratory animals, it is generally stated that the luteal cells are entirely of granulosal origin in the hamster, mouse and rabbit, but of dual origin in the rat and guinea-pig. However, unless (or until) it becomes possible to identify the type of steroidogenesis of individual cells and to distinguish them in this way if in fact they are of more than one functional type, the question of their origin is perhaps of little moment. In the rabbit ovary, where the interstitial tissue is particularly prominent, this tissue is the main source of 20α-hydroxy-progesterone, which it produces in larger amounts than the corpora lutea produce progesterone.[192] The 20α-hydroxy compound, however, is a much less active progestin than progesterone and it was shown, many years ago, that extirpation of the corpora lutea of the rabbit led to the termination of pregnancy.[79, 138, 158]

The luteal phase

After ovulation, when the discharged follicle has been converted to a corpus luteum, oestrogen production is reduced, and progesterone is secreted. The primary function of this hormone is to condition the uterus to accept and nourish the developing embryo, when ovulation has been accompanied by fertile mating. These 'progestational' changes (see p. 70 and p. 123) are marked by growth of the uterine glands and proliferation of the uterine stroma. In most species, they occur regardless of whether mating takes place, and whether it is fertile or not. Follicular maturation is inhibited until the corpora lutea cease to secrete progesterone and begin to regress. In a 'typical' case the corpora lutea of one cycle have almost disappeared by the time the next batch of follicles is mature. In the murine rodents, however, although ovulation occurs 'spontaneously' (that is, it does not depend on copulation—see p.74) and corpora lutea are formed from the discharged follicles, they do not, in unmated animals, secrete measurable amounts of progesterone. In these animals, therefore, succeeding 'crops' of follicles mature in rapid succession, so that the oestrous cycle is one of only four or five days (see p. 80). Several sets of corpora lutea may be found in the ovaries at one time, all but the most recent set being difficult to distinguish. The importance of this phenomenon lies in the usefulness of the rat and mouse as laboratory animals. The brief duration of the cycle is in itself an advantage in many investigations, and these species are in such general use in laboratories that the 4-day rhythm of the unmated rat is frequently referred to as the 'normal' cycle, although it is confined to a few species and to animals in captivity, for it is probable that wild rats or mice rarely experience oestrus without mating.

The usefulness of the 'normal' cycle of the rat in experimentation is such that an extraordinary creature, the 'cycling rat', has forced its way into the literature and has penetrated even to some of the most scholarly and fastidious journals. Its ovary normally contains several sets or 'generations' of corpora lutea because each lasts through several oestrous cycles. The 'inactive' corpus luteum of the unmated rat is, in fact, as long-lived as the 'active' (progesterone-secreting) corpus luteum of the pseudopregnant rat, but it does not hold back ovulation and the onset of oestrus. It reaches its maximum size in about 2 days in early dioestrus and only begins to regress about the middle of the following cycle.

In histological preparations of the rat ovary, it is usually possible to identify the most recent set of corpora lutea with tolerable certainty. The preceding set can often be distinguished, but beyond this it is impossible to assign the individual corpora to sets in this way. A single ovary may contain as many as 40 recognizable corpora lutea.

In contrast, there is never more than one set of corpora lutea in the hamster, although the unmated oestrous cycle is only of 4 days. They show signs of regression by mid-cycle and then regress very rapidly.

Progesterone synthesis

The principal steroid precursor stored in the ovary is cholesterol. It seems clear that the ovary can receive cholesterol from the bloodstream and that it can, also, synthesize it from precursor compounds. Its conversion to progesterone has been demonstrated *in vitro* and *in vivo*. The probable course of events is that cholesterol first undergoes hydroxylation (in two stages). Cleavage of the side chain then yields isocaproic aldehyde and pregnenolone, and the latter is converted to progesterone by the action of the enzyme Δ^5-3β-hydroxysteroid dehydrogenase. This enzyme is detectable by histochemical methods as described elsewhere (p. 136). It is apparently universally present in steroid-secreting cells, with the possible exception of the rabbit corpus luteum. In this tissue, the enzyme was not demonstrable by the histochemical technique which in the same hands gave a positive result in luteal tissue from many other species, and in the interstitial tissue of the same rabbit ovaries.[326] The authors concluded that 'The corpora lutea of the rabbit may have a pathway for steroid hormone biosynthesis that differs from that in the interstitial tissue and adrenal cortex of the same animal. However, the absence of activity may refer to the histochemical method only'. The paper ends, like many others, with the observation that the subject requires further investigation.

Further work showed that the rabbit ovary appeared to produce its progestin by means of its interstitial tissue,[326] although, as we have seen, the corpora lutea had long been known to be essential for the maintenance of pregnancy in this species. 'In the rabbit, all Δ^5-3β-hydroxysteroid dehydrogenase activity is located in the interstitial tissue whether the rabbit is pregnant or not. In pregnancy there is marked hypertrophy of the interstitial elements. There are also large corpora lutea, but in these we have been unable to demonstrate enzymic activity histochemically under any of the conditions applied. At the present time we are completely unable to reconcile the absence of Δ^5-3β-hydroxysteroid dehydrogenase activity with Fraenkel's finding, but it is compatible with the observation that ovaries without corpora lutea secreted as much progestin per mg tissue as did those with corpora lutea and follicles',[191] The word 'progestin' of course embraces any progestagen, not only progesterone.

Subsequent investigations have inclined to the explanation that the corpus luteum of the rabbit converts pregnenolone to progesterone and secretes it as such, by the same biosynthetic pathway as demonstrated in other species. 'Rabbit corpus luteum homogenates incubated *in vitro* have been found capable of converting pregnenolone to progesterone'.[16] This, it is implied, must involve hydroxysteroid dehydrogenase, and the failure to demonstrate the enzyme histochemically is probably attributable to 'some artifact of the histochemical procedure which is peculiar to the rabbit corpus luteum'. This is now generally accepted, as sensitive chemical assays have shown that the 'progestin' produced by the interstitial tissue is in the form of 20α-progesterone (20α-hydroxy pregn-4-en-one). This steroid is much less effective than progesterone itself (pregn-4-ene 3,20 dione), both in sensitizing the uterine mucosa and in the subsequent maintenance of pregnancy. The corpora lutea, on the other hand, produce progesterone as such. Thus, although the interstitial tissue produces steroid in such quantities as to obscure the difference in total progestin output between an ovary with corpora lutea and one without, the effective progestagenic agent is mainly produced by the corpora lutea, and this explains their indispensibility in pregnancy.

These observations imply that the conversion of pregnenolone to progesterone in the rabbit's luteal tissue is not detected by the histochemical technique that has proved effective elsewhere, even in the same ovary. This in turn enjoins caution in interpreting a negative result when this test is applied. More importantly, we are left with the problem of the possible function of the 20α-progesterone produced in such quantity by the rabbit interstitial tissue, which undergoes marked hyperplasia to produce it. It may be unforgivably teleological to demand a 'use' or 'purpose' for any or every biological phenomenon, but the attitude of 'I see it is there, I wonder what it does' has led to many productive enquiries.

Because 20α-progesterone is so 'weak' a progestagen compared with progesterone, it has been suggested that it may function as an 'antiprogesterone' by competing with the more effective compound for the binding sites involved in its action within the target tissue. This could conceivably be a means of terminating pregnancy when, as in the rat, the plasma level of the 20α-compound rises suddenly as the progesterone level falls. It may, for instance, compete for the binding site on the protein molecule that protects the steroid from elimination in its passage through the liver (see p. 133).

Slices of luteal tissue maintained *in vitro* have been used to investigate the action of various substances on progesterone synthesis (see references to Schomberg's experiments, p. 112). The technique has obvious limitations but, equally, it has advantages. In particular, it makes possible the exclusion of substances other than those under test, in investigating the effect of hormones or drugs on the stimulation or suppression of progesterone synthesis. Largely by the use of this type of experiment, it has been shown that steroidogenesis depends on LH stimulation in some species if not in all. Savard, Marsh and Rice, in a comprehensive review of 'Gonadotrophins and ovarian steroidogenesis',[333] suggested that 'two types of corpora may exist; those which produce only progestins, consisting of progesterone and one (or both) of the epimeric 20-hydroxy-4-pregnen-e-ones, and those which produce both progestins and oestrogens (plus other steroid substances, 17-hydroxy progesterone, 4-androstenedione, etc.)'. They quote the corpus luteum of the cow as an example of the former category, and the human corpus luteum as an example of the second type. The corpus luteum of the mare is probably like that of the cow in producing no oestrogen.

Experiments using this same technique of incubating luteal tissue slices in a medium containing test substances have produced evidence that prolactin, which promotes luteal activity in the rat, has no effect on steroidogenesis in women, or in the monkey, sow, ewe, guinea-pig or rabbit. As described elsewhere (p. 99) other evidence suggests that prolactin may be involved in progesterone secretion in the ewe, and some experiments in my own laboratory suggest that it may play some part in luteal activity in the guinea-pig*. It may well be that prolactin affects not the synthesis of progesterone but its release or, more probably in the case of the rat, the survival of the corpus luteum as an active gland. We are, yet again, in danger of being confused by the varied, and changing, use of terms. It would be useful to restrict the meaning of 'luteotrop(h)ic' to 'contributing to the maintenance of the corpus luteum as a histological entity'. Substances which promote steroid production could be referred to as being steroidogenic (except that this is used of the tissue itself) but we would still lack a collective term for substances (if there are any) that promote or control the release of hormone from the (luteal) tissue. Up to now the term 'luteotrop(h)in' has not been limited in this way, because luteal survival and function could not be clearly distinguished in the earlier investigations. It behoves the reader, therefore, to consider the sense in which this term is being used, whenever a distinction between the survival and the function of the corpus luteum is of moment.

Ovarian ultrastructure

This term 'ultrastructure' has come into use in biological contexts to indicate cell structure and histological detail as revealed by electron microscopy. The technique had not long been available before it was applied to the germ cells, both male and female; students of fertilization were eager to study the spermatozoon and the ovum and, especially, the zona pellucida which the former has to penetrate to reach the latter. Then the follicle cells, the luteal tissue, and the interstitial tissue were studied in this way, and a fairly comprehensive picture of ovarian ultrastructure has emerged. The different types of tissue have naturally been

*Illingworth and Perry, 1971 (*J. Endocr.*, in press): Luteal growth in hypophysectomized guinea-pigs is increased by prolactin injections, and corpora lutea grow more quickly after pituitary stalk section than after hypophysectomy.

studied in a considerable variety of species, and the ultrastructure of each type has proved to be fairly similar throughout.

The surface of the oocyte is produced into numerous processes or microvilli, which penetrate radially for varying distances into the zona pellucida. Rather more prominent processes penetrate the zona from the follicle cells, and extend through it to make contact with the egg membrane. These 'processes' are evidently formed as the zona material is secreted, principally if not entirely from the follicle cells. As the zona thickens the follicle cell wall is pushed away from the egg surface except at restricted points, where contact is maintained. When the egg is mature, and before it is shed, most of the follicle cell processes are retracted, but a 'cumulus' of follicle cells remains around the egg when it is liberated. Whether these cells continue to serve any function with respect to the ovum, in its passage into the fallopian tube, seems not to have been determined. It is easy to imagine that they have a passive or mechanical function in that they provide bulk, and may therefore facilitate the transporting action of the fimbrial cilia. This, however, is a speculative interpretation and the real situation may be quite different. The cumulus cells are soon dispersed, but whether they 'die' at this time, or whether their death contributes to their dispersal, is not clear. Presumably, however, they retain in some degree the relation to each other as cumulus cells that they possess while in the follicle. Electron micrographs show that even in the early stages of oocyte development the innermost follicle cells are only loosely bound, and they become progressively more openly spaced as the oocyte grows.

The cytoplasm of the oocyte itself contains a number of organelles, including mitochondria, lysosomes and in many cases dense clusters of membranous vesicles ('ergastoplasmic cisternae'). The Golgi apparatus is prominent and there are paracrystalline bodies in many oocytes. Lamellar structures are prominent in human oocytes. Most of the organelles are found clustered near the nucleus in the early stages of oocyte development (Plate II). They form a relatively dense crescentic mass in some optical microscope preparations. Seen thus, it was known as the 'Balbiani' vitelline body. In this case, as with many other cytoplasmic structures, the electron microscope has proved the reality of what had previously been suspected to be 'artifacts'. This aggregation of mitochondria and other membrane structures in a paranuclear position gives the oocyte a degree of polarity. It is quite probable, but by no means certain, that this structural polarity is of functional significance.

Later, the oocytes of many species are characterized by 'cortical granules' arranged peripherally below the vitelline membrane. Long projections of the follicle cells have been described, that penetrate right through the zona pellucida and into deep tubular invaginations of the limiting membrane of the oocyte, so that they run centripetally within the oocyte cytoplasm and 'may reach deep into the oocyte cytoplasm to terminate a few microns from the oocyte nucleus'.[23] Apart from these infrequent processes, however, electron micrographs of the zona pellucida, in human and other species, give the impression that the zona pellucida almost, or completely, separates the oocyte from the follicle cells. As the zona thickens, the follicle cells, by whose secretion it is presumably formed, retreat away from the oocyte, leaving many thin tubular cytoplasmic processes entrapped within it. The cytoplasmic projections on the surface of the oocyte are much shorter, and take the form of rather stumpy microvilli, penetrating only a little way into the zona pellucida.

The 'zona' is, characteristically, a tough and resilient structure, sufficiently elastic to recover immediately from deformation when handled under a low-power microscope. The electron microscope, however, does not reveal any morphological detail in its ground substance. It is composed mainly of mucopolysaccharide, which is of medium electron density, and it is without fibrils or crystalline structures, so that it appears a uniform grey in electron micrographs.

Mention has already been made of the existence, and evident importance, of the membrane on which the follicular epithelium (in the ripe follicle the membrana granulosa) stands. It too is spherical, resilient and continuous. This membrane consists of the true basement membrane of the epithelium, a thin 'structureless' layer that extends beneath the epithelium and is continuous across the cell junctions, and a relatively thick collagenous layer underlying the basement membrane, and separating it, and its epithelium, from the connective tissue of the theca interna. Thus, in the rabbit follicle, at a stage when the granulosa is three to five cells thick, the outermost layer 'lies over a thin basement membrane. It has now become separated from the surrounding connective tissue cells by a regular space, $2-3\mu$ wide, which is partially occupied by bundles of collagen fibrils'.[397] The authors illustrate this arrangement and also present a very clear electron micrograph of the primordial follicle wall, showing the same 'basement membrane' extending around the single layer of flattened follicular cells. In the later stages of follicular development it is evident that collagen fibrils, and the matrix filling the 'space' in which they lie, become organised into a sheath of very regular thickness, and a large follicle can be dissected free of surrounding tissue by virtue of this resilient 'skin'. Although both collagen and matrix are clearly contributed by the cells of the theca interna, they form, together with the true basement membrane, a capsule into which the thecal cells do not penetrate, and through which no blood capillaries penetrate until very near the time of ovulation.

Electron microscopy is not well adapted for the study of intra-nuclear structures, and although the formation of pronuclei, and stages in meiosis, have been studied, relatively little has been added to what optical microscopy had already demonstrated. The first polar body is a miniature replica of the oocyte, lenticular in shape and rather less regular in outline than its large sister cell. If it divides again its daughter cells have a rather simpler cytoplasmic structure.

With the approach of ovulation, the granulosa cells undergo some changes in ultrastructure. In particular, there is considerable increase in the smooth endoplasmic reticulum. Bjersing has suggested that this is formed by proliferation of the granular or rough endoplasmic reticulum, as he observed direct continuity between the two varieties of membrane.[28] Some lipid droplets are apparently localised in cavities or pockets in the endoplasmic reticulum, presenting an appearance which, like that of the proliferative smooth endoplasmic reticulum, is reminiscent of the luteal cells into which these granulosa cells will shortly be converted. It has been suggested, on the basis of experimental treatment and subsequent electron microscopy, that the membrane proliferation is stimulated by FSH, and the formation of lipid granules by LH. These changes are detectable by electron microscopy, before the cells enlarge and assume the characteristic 'luteal' appearance seen later in optical microscopy.

The corpus luteum

The 'luteal' cell has a characteristic appearance under optical microscopy, and the general similarity from species to species is also evident in accounts of its appearance in electron microscopy. The cytoplasm of these large cells in the fully active corpus luteum is crowded with whorls of smooth endoplasmic reticulum and groups or clumps of subspherical lipid droplets (Plate IV). Mitochondria are interspersed among these droplets (some authors refer to them as lipid granules) and sometimes appear to be crowded out into peripheral regions of the cell. It seems to be characteristic of the tissue that the lipid droplets are of two kinds, distinguished by different electron opacity.

The corpus luteum is provided with a well-developed vascular and lymphatic drainage system. The theca of the mature follicle carries a network of vessels, but, as already described, they do not penetrate the basement membrane (membrana propria) of the lining (granulosal) epithelium until ovulation is imminent, or until it has occurred. But as the granulosa cells enlarge and 'luteinize', connective tissue elements ramify among them and rapidly organise them into radially-disposed columns. Capillaries run in the connective tissue and soon come to lie in intimate association with the columns of luteal cells. They do not, however, make direct contact with the secreting cells. Between the luteal cells, and between them and the endothelium lining the capillaries, is a continuous and complex 'sub-endothelial' or 'pericapillary space'. This 'space' actually communicates with the lumen of the capillary at some points, through 'gaps' in the endothelium. In my material, derived from pregnant guinea-pigs, these gaps, when they occur, lie between adjoining endothelial cells, and the junction is usually marked by a fibrillar structure resembling the 'terminal bar' found in various tissues. It is difficult to imagine, on the basis of this material, that these infrequent openings constitute the main pathway by means of which material passes into the blood system. They may well serve for the passage of certain molecules, but the traffic must, one would suppose, involve the very numerous and complex cell processes that ramify in the sub-endothelial 'space' from the luteal cells on one side of the channel and, to a lesser extent, from the endothelial cells on the other. These surface structures are shown, for example, in Plate IV. Gaps in the endothelium of capillaries within the corpus luteum appear much more numerous and obvious in the sheep material described in more detail below. Whether the difference is between species, stage of cycle or techniques, is not yet clear. In examining and interpreting any electron micrograph or, for that matter, any photograph or photomicrograph of fixed material, one must remember the distortion that inevitably accompanies fixation. In this case, one cannot know the relative dimensions of the 'space' and the cytoplasmic processes during life. It is particularly tantalising that one cannot tell whether these processes are semi-permanent or whether they grow and diminish, swell and subside, from moment to moment. Electron micrographs of blood capillaries show the endothelial cells with relatively enormous nuclei that protrude into the lumen and sometimes nearly occlude it, especially in tissue fixed by immersion and not by perfusion. It is evident that this picture is caused by the shrinkage of the lumen under dehydration during processing of the tissue, and tissue such as that of the corpus luteum would be much less dense if one could preserve and observe it without distortion.

Within the cytoplasm of the luteal cell after ovulation, there is a very marked increase of ('smooth') endoplasmic reticulum (s.e.r.), often arranged in the complex 'whorls' that appear to be characteristic of steroid-producing tissue. Numerous free ribosomes are found in the cells with well-developed s.e.r. The conspicuous whorls often, perhaps usually, envelop a number of lipid granules, as shown in Plate IV. The simplest interpretation of the relations between the intra-cellular membranes and the steroid-secreting function of the cell is to imagine the lipid product to be formed in the laminate channels in the e.r., transported to wider lacunae and to Golgi channels, and collected into spherical capsules, which are then cut off from the foliate membrane structure and appear as free 'droplets' or granules. They are not, however, released from the surface of the cell as encapsulated globules, for there are no lipid granules in the subendothelial space or in the capillaries beyond. Perhaps more probably, the large lipid granules may be supposed to constitute the material from which the steroid products are synthesized (see below, p. 42).

It has been shown, by direct measurement, that progesterone is drained from the corpus luteum by the lymphatic system as well as by the blood vessels. Some investigators, indeed, have suggested that the lymphatic drainage is more important, in this respect, than the vascular drainage. But although the concentration of steroid hormone may be as high, or higher, in the lymphatic vessels as in the ovarian vein, the rate of flow is very much greater in the latter. Furthermore, the lymph vessels do not penetrate among the luteal cells to the same extent as the blood capillaries. This is very clear from the excellent account of the lymphatic drainage of the ovary based on that of the sheep and the rat, by Morris and Sass.[263] The lymphatics within the ovary grow and subside with the development and regression of follicles and corpora lutea, and it has been shown that there is no measurable flow in the lymph vessels draining the ovary except when large follicles, or corpora lutea, are present. The 'sub-endothelial' or 'pericapillary' spaces already referred to are not lymph vessels. They are not lined by an endothelium and indeed they are not truly 'spaces' but rather very attenuated connective tissue, presumably a form of gel through which soluble substances can diffuse freely. These 'spaces' also ramify between the clumps of luteal cells and anything secreted by the latter must first enter the 'space'. The products of secretion are perhaps transported directly from the 'space' immediately outside the luteal cell into the lymphatic drainage, or they may enter the bloodstream directly through the 'gaps' in the capillary walls, or be passed through the endothelium into the capillaries.

Morris and Sass showed that 'In the ewe there was a big difference between the lymphatics of inactive ovaries and ovaries with functional corpora lutea. Coincident with ovulation and the formation of a corpus luteum, lymph flow from the ovary increased manifold. The high rate of lymph flow and the high protein content in this lymph suggested that the ovarian capillary bed, or at least part of it, had become very permeable'. Increased capillary permeability just before ovulation had been demonstrated in rabbits many years earlier, and attributed to the action of oestrogen. It had also been found that whereas the concentration of progesterone in the ovarian lymph is high during the luteal phase of the cycle (and in pregnancy) that of oestrogen remains low even during the follicular phase. It was therefore concluded that 'The fact that the capillaries of the corpus luteum appear to be the most permeable, and that the luteal cells are responsible for the production of progesterone, suggests an association between the synthesis and secretion of progesterone and an increase in capillary permeability'. The increase in capillary permeability is presumably due in part to relaxation of the 'gaps' between endothelial cells. As the increased permeability lasts throughout the luteal phase (i.e., for the functional life of the corpus luteum) it seems reasonable to suppose that it is caused by progesterone rather than oestrogen. The fact that it begins before ovulation does not necessarily preclude this, because pre-ovulatory progesterone secretion, presumably by cells of the theca interna, has been demonstrated in a number of species.

A beautifully illustrated account of the development and definitive structure of the lutein cell is given for the rabbit by Blanchette.[30] Her electron micrographs show very clearly the extraordinarily massive whorl formation of s.e.r., and the so-called 'myelin-figures' which form as these whorls collapse when the cell shrivels and dies. These myelin figures are presumably associated with the 'fatty degeneration' that accompanies luteal regression as seen by optical microscopy. For it is notable that the regressing corpus luteum stains as heavily with lipid stains as does the mid-phase cell, but during regression the fat globules are fewer and larger. There are only one or two per cell, whereas the active cell contains a large number of very much smaller 'globules'.

A more detailed consideration of the functional significance, rather than the morphological appearance, of the structures characteristic of the corpus luteum, is provided by Bjersing in his work on the pig. He noted that the increasing quantity of s.e.r. and its prominence in 'whorls' in the luteal cell were paralleled by an increase in the average volume of the individual mitochondria. The Golgi apparatus became prominent early in the luteal phase. 'Cytoplasmic bodies, morphologically consistent with lysosomes' were found in increasing numbers in the later stages of the cycle. Bjersing agreed with Deane's hypothesis, based on histochemical evidence, that the lipid droplets in steroid-producing cells represent 'stores of potential precursor materials that may be converted into steroid hormones when the proper stimulus occurs'.[89] Bjersing considered that the lipid droplets form within 'dilated cavities of the endoplasmic reticulum' and become bound up in the whorls of s.e.r. as these are formed by the accumulating membrane. He argued that the droplets probably include cholesterol among their constituents, and suggested that 'One factor in the diminution of droplets during dioestrus, when the progesterone synthesis is maximal, may be the migration of cholesterol to the mitochondria, where the cleavage of its side chain presumably takes place'. This, he postulated, may account for the fact that free lipid droplets are sometimes absent from cells that are known to be actively synthesizing steroids: in effect, he suggests, conversion may sometimes outstrip the supply of precursor material so that the supply is used immediately on arrival, and no stores accumulate.

Lysosomes and luteal regression

The regression of the corpus luteum is not merely the elimination of used-up and expendable material. It is crucial to the control of ovarian periodicity, to which a separate chapter will be devoted (p. 89). The cytology of the process therefore assumes a special importance. Near the end of the cycle, the corpus luteum obviously regresses, becoming smaller and undergoing rapid changes that soon lead to its virtual disappearance in many species. Bjersing found that the active secretion of progesterone ceased in the pig before these changes were detectable either grossly or by optical microscopy or by means of the electron microscope. This was also reported to be the case in the sheep[90] and further work on the corpus luteum of this species indicated that involution of the luteal cells is brought about by the release of autolytic enzymes from the lysosomes.[102]

The term 'lysosome' covers a morphologically heterogeneous group of cytoplasmic particles that contain acid hydrolases and serve as an intracellular digestive system. Their identification as organelles specifically concerned with the control of 'destructive' enzymes is largely owing to C. de Duve and his collaborators in Louvain. Their origin and function was briefly summarized as follows: 'It has become apparent . . . that the relatively primitive activity of intracellular digestion has been adapted in specialised cells of multicellular organisms to subserve a variety of functions such as defence, absorption, differentiation and cell involution'.[364] It is the last-named function that is involved in luteolysis, and the connection with the 'primitive activity of intracellular digestion' (phagocytosis) perhaps justifies a brief digression.

These particles are intermediate in density and size between mitochondria and microsomes, and early fractionation methods included them sometimes with the former and sometimes with the latter. By improved techniques the Louvain group were able to isolate a well defined fraction, the hydrolase activity of which could be enhanced by osmotic change, mechanical disturbance, or treatment with detergent.[383] They eventually concluded that they were 'dealing with a new type of cellular organelle delimited by a single membrane, responding to osmotic shock, and containing mainly acid hydrolases. The new granule born in the test-tube of the biochemist was named lysosome before the morphologists had glanced at it'. This quotation is from the chairman's introductory remarks at a meeting during which de Duve himself read a paper outlining the postulated phylogenetic history of the lysosome, and attributing the basis of the idea to Metchnikoff who, in the closing years of the nineteenth century, demonstrated the role of leucocytes and macrophages in defence against bacterial invasion. Previously, the observation that white blood cells congregate at an infection site and there die in great numbers 'obvious prey to the invading bacteria' had been interpreted as showing that bacteria thrive on leucocytes. But Metchnikoff saw these cells as (in de Duve's words) 'the direct evolutionary descendants of amoebae and paramecia. In them, he saw the primitive function of intracellular digestion preserved throughout evolution, long after its original nutritive purpose had disappeared, to become an essential factor in the natural immunity of higher animals'.[97] He coined the term 'cytase' to define the soluble ferment which, he maintained, was involved in intracellular digestion; he realised that the process took place in an acid medium, and that the living cell must possess some kind of barrier to protect it from its own cytases. 'In fact', writes de Duve, 'there is little doubt that if the techniques of cytology and biochemistry had been somewhat more advanced than they were, Metchnikoff would have discovered the lysosomes'. He adds: 'Now that electron microscopy has revealed the widespread distribution of pinocytosis and related phenomena . . . it is becoming evident that Metchnikoff's evolutionary concept is of much wider and more universal application than he himself probably was aware of. As our present evidence indicates, the ability to engulf foreign objects or molecules and to digest the engulfed material has been retained in most, if not all, animal cells. Immunity is but one of the functions fulfilled by this basic process, which has become adapted in different cells to an astonishing variety of purposes'.[97]

In the work on the sheep, already referred to, corpora lutea were obtained at different stages of the cycle, and a sub-cellular fraction was prepared that appeared to contain a substantial proportion of lysosomes. Those from corpora lutea taken near the end of the cycle were more easily disrupted than those from younger corpora lutea or from corpora lutea of the same age during pregnancy. The general concept of lysosomal function is that these bodies are membranous sacs containing autolytic (self-destructive) cell enzymes. The cell's activity, and its existence as an organised system, can be brought to an end by any process or factor that acts on the lysosomal membrane in such a way as to release the contained enzymes. In this work it was shown that the active corpus luteum of the sheep contains acid phosphatase and acid protease in significant amounts. First, histochemical methods indicated that the activity associated with acid phosphatase was confined to well-defined particles except at the end of the cycle. Next, it was shown that while most of the acid phosphatase and acid protease activity could be isolated by sedimentation from a sucrose homogenate, the activity was not manifested unless the sedimented particles were disrupted, for example by treatment with detergent. It was possible to measure the total enzymic activity (of these two enzymes) and the 'free' activity. The difference was taken to represent the 'bound' activity—that is, the enzymes locked up, as it were, in the lysosomes. The proportion of the total activity that was 'free' at any one time was taken to indicate the extent to which the lysosomes were disrupted by the homogenisation process. This proportion remained almost constant, little over one third of the total activity being 'free', until the 15th day of the 16-day cycle, when it suddenly rose to nearly two-thirds. No such rise in the proportion of 'free' enzyme activity was observed in the pregnant animal.

The lysosomes were tentatively identified as the electron dense bodies, 0·3 to 0·5μ in diameter, seen in earlier investigations.[90] They were thought to decrease in number during the last day or two of the cycle. The authors put forward the suggestion that the functional 'life' of the corpus luteum is perhaps determined by a specific substance, of uterine origin, causing the sudden increase in lysosome fragility and thus releasing autolytic enzymes within the luteal cells. If such a substance exists, it would fulfil the requirements for the postulated 'luteolysin', discussed elsewhere. As the authors clearly recognized, however, the solution to this problem is by no means cut and dried. The idea of a specific lysosomal 'trigger' accords with the fact that regression of the corpus luteum begins with changes in the luteal cells themselves, and the sudden increase in lysosomal fragility coincides with the dramatic fall in progesterone secretion, at least in the sheep, but the role of the lysosomes may be a secondary rather than the primary one in the sequence of events within the luteal cell.

An alternative but not wholly dissimilar explanation of the mechanism of luteal regression arose from work on the rat.[24] Here again lysosomal breakage was considered to be involved in luteal regression, but it was thought that most of the lysosomal enzymes involved were carried to the corpus luteum in macrophages that were found to invade the ovary in the later stages of luteal activity. This suggestion conveniently provides a possible means of transport for the lytic enzymes, but neither the origin of the macrophages nor their route has been demonstrated and their role remains hypothetical.

We have seen how rapidly the ruptured follicle (that is, the newly formed corpus luteum) acquires an elaborate and extensive vascular supply. This supply is even more dramatically reduced when the corpus luteum regresses. In the pig, for example, the corpora lutea shrink and become quite white in the last few days of the cycle. The regressed body is, of course, generally known as a corpus albicans as distinct from a corpus luteum—terms that simply describe the change from a yellow body to a white body. If the primary factor in luteal regression is the collapse of the luteal cells, then the vascular shut-down must be in response to a suddenly diminished demand on the part of the tissue concerned. It has been shown, however, in a different context, that a lowered intracellular oxygen tension is a principal cause of lysosomal 'rupture', and it may well be that a reduction in the vascular supply is involved in luteolysis, since this could presumably lead to the release of the autolytic enzymes from the luteal cell lysosomes.

An indirect role for the luteal cell lysosomes has been suggested: 'Some other element of the lutein cell may initially be affected, the changes in lysosomal function thus being of a secondary though nevertheless important nature'.[102] There is evidence that the actual cessation of steroid synthesis may lead to autophagic changes in the cell.[364] This is a reference to work by Smith and Farquhar who were concerned with the role of lysosomes in the anterior pituitary. This postulated a variety of lysosomal types, one of which, they suggested, has the function of 'digesting' surplus product when secretion is stopped. This account is notable for its full and precise description of the techniques involved and for the justification of their elaborate care in preparative method that is provided by the high standard of their electron micrographs of secretory tissue.

In contrast to the observations quoted above, to the effect that in sheep and in pigs corpus luteum function, as indicated by progesterone secretion, ceased before the cells underwent any obvious change, it has been suggested that the human corpus luteum at term, which looks 'regressed' under optical microscopy, may in fact possess considerable amounts of smooth endoplasmic reticulum; the cytoplasmic organelle systems appear intact.[148] As these cells cease to secrete progesterone in the first trimester of pregnancy, the membrane systems must either retain their structural integrity quite 'unnecessarily', or be put to some other use.

The ultrastructure of 'active' luteal tissue has been referred to in the preceding account, with little reference to its origin, beyond the name of the species concerned. This is justified by the remarkable similarity of the tissue throughout the range of physiological states in which it has been studied. There is, too, relatively little variation from species to species. The same is largely true of the optical microscopy of the luteal cell, except that the cell boundaries appear more prominent in some species than in others.

3: The Basic Elements of the Ovarian Cycle

Outline of the endocrine mechanism

In studying the literature concerning the ovarian cycle, one is impressed by the antiquity of some of the observations and the penetration of the conclusions drawn from them, and also by the obstinacy of many important features in resisting elucidation in spite of the weight of investigation to which they have been subjected in recent years. Various parts of the mechanism are understood in some detail, others hardly at all, and the attempt to provide an outline of the whole is further complicated by differences between species. Such differences, which often seem to bear little relation to the phylogenetic relationships of the groups, extend to all but the most fundamental of the processes involved.

The phenomena associated with reproduction are almost universally cyclic in nature. The succession of generations is itself an example of this, and in each generation the successive breeding seasons are usually governed by the annual climatic cycle, which imposes its rhythm on the organism. In the female mammal there exists an internal rhythm that governs the regular periodic release of eggs from the ovary; the mechanisms that govern and modify the ovarian cycle, and the phenomena associated with them in the mammals, form the subject of this book. If one is asked 'What can be said of the ovarian cycle that will apply to all mammals?' one is tempted to reply that there is precious little. A common framework does exist, however, and it may be described briefly as follows.

Figure 2 illustrates diagrammatically the principal relevant endocrine pathways as they are generally accepted at present. Ovarian follicles, and the oocytes within them, mature under the influence of *follicle-stimulating hormone* (FSH). This is a gonadotrophin secreted

47

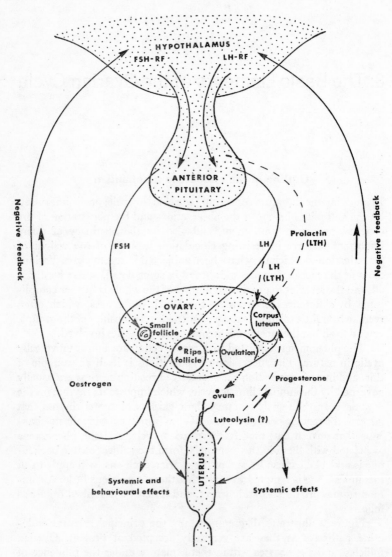

Fig. 2. The principal endocrine pathways involved in the control of ovarian function in a mammal.

into the bloodstream by the anterior pituitary gland in response to a 'releasing factor' (FSH-RF) which travels down the pituitary stalk from the hypothalamus. The developing follicle secretes an *oestrogen* (mainly *oestradiol*) which affects the reproductive tract and (except in the higher primates) induces oestrus or 'heat', the condition in which the female is sexually receptive. When the follicle is 'ripe' it ruptures, and expels its oocyte, under the influence of another gonadotrophin, *luteinizing hormone* (LH), which is released from the anterior pituitary gland by a second releasing factor (LH-RF). The ruptured follicle is rapidly converted into a solid glandular body, the corpus luteum, by luteinization of its lining cells; hence the name given to the gonadotrophin concerned. The corpus luteum secretes a *progestagen*, predominantly *progesterone*. This hormone brings about 'progestational' changes in the uterus, in readiness for the implantation of an embryo. After a time, the corpus luteum will regress, more follicles will ripen, and oestrogen will again predominate over progesterone in governing the changes in the accessory sex organs and in sexual behaviour. The cycle may be thought of, therefore, in terms of an alternation in the dominance of these two ovarian hormones, both of them steroids. This alternation is controlled by a 'feedback' mechanism, since both oestrogen and progesterone, when they reach a sufficient concentration, inhibit the secretion of their respective pituitary stimulators. The effect is operated through the hypothalamic releasing factor in each case, as indicated in the diagram. In the case of oestrogen, very low concentrations have a 'positive feedback' effect; the control of its secretion is thus very delicately balanced and can achieve a rapid build-up and a sharp cut-off.

The corpus luteum appears to have an intrinsic viability which differs among different species, but its maintenance usually appears to depend on a pituitary *luteotrophic hormone* (LTH). In some species, notably the rat, this function appears to be fulfilled by *prolactin*, the hormone which is associated with lactation, as its name implies, and with a great variety of other functions as well.[312] In other species, probably the majority, the luteotrophic function is served by LH, so that in these species the formation of the corpus luteum and its continued function both depend on the same hormone. In many species the uterus of the non-pregnant animal plays a part in the regression of the corpus luteum. This may be a positive 'luteolytic' activity, as distinct from the negative effect of withdrawing luteotrophin—'killing' rather than merely 'failing to keep alive'. The factors responsible for this effect have not yet been identified, but the evidence is such that the existence of a distinct uterine hormone, '*luteolysin*', has been postulated.

The nature of the hormones

Pituitary aberrations such as Fröhlich's syndrome, and Cushing's syndrome, have long been known to involve gonadal malfunction, but the existence of pituitary gonadotrophin was not demonstrated until 1927, when it was described, independently but almost simultaneously, by Smith and Engle[363] in America and by Zondek and Aschheim[398] in Germany. Some years elapsed before it was generally accepted that two distinct substances, FSH and LH, were involved, and they have not even yet been completely isolated and identified.[59] It seems safe to predict, however, that two distinct entities will emerge, the properties of each having become more distinct with successively more highly purified preparations.

FSH and LH are protein hormones; the ovarian oestrogens, and progesterone, on the other hand, are steroid compounds, derived from cholesterol. They have been synthesized and are available in pure form. They are secreted not only by the ovary but also by the testis and by the adrenal gland and, in some species, by the placenta. Oestrogenic and progestagenic effects are not confined to the steroids, however. Certain plants contain weakly oestrogenic compounds, such as genistein, which, like the potent and widely used synthetic oestrogen stilboestrol, is non-steroidal. Non-steroidal progestagens, such as 'methallibure' (I.C.I. 33828) have been developed in the search for contraceptive agents suitable for oral administration.

The exact nature of the hypothalamic 'releasing factors' remains obscure. They are probably not large molecules, and therefore not proteins, but the process of identifying and defining them has so far reached only the stage of the extraction of minute amounts of them from vast quantities of hypothalamic tissue, as described elsewhere (p. 59). In 1932, when a similar 'availability' problem beset research on oestrogen 'standards', a French scientist, Dr. A. Girard, diffidently offered to a group who met under League of Nations auspices, 20g of what had hitherto been available in milligram quantities. He had in fact discovered a method of tapping a readily available source of oestrogens, the urine of the mare from the seventh to the tenth month of pregnancy. A similar 'princely contribution', as A. S. Parkes called it, would doubtless be welcomed by the investigators of releasing factors, but the minute quantities in which they are produced makes anything of the sort improbable.

Gonadotrophins, and their metabolic products, are excreted in the urine in man, and a great deal of work has been done on them by extraction from this source. They are not excreted in this way by non-primate mammals.

Mode of action of hormones on target tissue

Until we know far more about the mechanism of the response of the target tissue to the controlling hormone 'at the cellular level' (to use a phrase much in evidence at the present time) we shall not understand the mode of action of any hormone in any given circumstances. Cyclic variations in the concentration of hormones are observed in the secreting glands and in the circulating plasma, but it is not always clear that a higher concentration means a more pronounced response. Nor is it clear, in many cases, whether the hormone becomes involved in the response, and is metabolized within the target tissue, or whether it exerts its effect *en passant* and is eliminated elsewhere.

An important step forward has been made in this direction in recent years by the discovery of the role of a cytoplasmic component of very general distribution in tissues—cyclic AMP and its activator, known as adenyl cyclase. This advance stems from the work of Sutherland and his collaborators at Vanderbilt University in Tennessee.[370, 375] It derives from the earlier demonstration that the effect of adrenaline, in increasing the rate of conversion of glycogen to sugar in the liver, was brought about by its increasing the activity of the enzyme phosphorylase, which catalyses the conversion. It was found that phosphorylase exists in two forms, one very much more active than the other. The interconversion of these two forms is controlled by the balance between an activating and a de-activating enzyme, and the action of adrenaline is upon these. The existence of a further intermediary was indicated by the fact that, although all the elements mentioned so far are soluble, and although the activation of phosphorylase proceeded in tissue homogenates, it did not take place in a homogenate from which all the particulate material had been removed by centrifugation. When the particulate and soluble fractions were re-united, however, the activity was restored. The factor concerned proved to be adenosine-3′,5′-monophosphate, referred to as 'cyclic AMP', residing in the ribosomes of the cell. It is a derivative of adenosine triphosphate (ATP), the universal energy transmitter of living cells. The existence of an enzyme (adenyl cyclase) with the function of converting ATP to cyclic AMP was postulated, and such an enzyme was subsequently found to reside on or within the plasma membrane, the outer boundary of the cell. Underwood describes this fact as 'teleologically satisfying' because it implies that the adenyl cyclase lies right at the surface bathed by the fluid that carries the hormone to the cell, and it is perhaps not necessary to suppose that the hormone even enters the cell.

More recently, cyclic AMP has been shown to be involved in the target cell's response to a number of hormones, including LH. It has been shown that the function of cyclic AMP in steroidogenesis is to increase the conversion of cholesterol to pregnenolone.

It is postulated that although adenyl cyclase always has the function of converting ATP to cyclic AMP, and thereby releasing the special function associated with the particular cell, the enzyme is not identical from tissue to tissue. The adenyl cyclase involved will be specific to the hormone to which the cell is designed to respond.

Cyclic AMP is removed, very quickly, by another enzyme within the cell, phosphodiesterase, which converts cyclic AMP to 5'-AMP and so limits the hormonal effect.

It is probable that there exist other 'second messengers', as cyclic AMP has been called—the hormone being the 'first messenger' in each case—and it is evident that the nature of the 'receptor site' on the surface of the target cell, and that of the reaction between the hormone and the adenyl cyclase, are yet to be elucidated.

Ovarian response to gonadotrophin

It has been suggested that the ovary acquires the capacity to respond to gonadotrophin in the course of its own development, independently of extragonadal factors. The ovaries of prepubertal rats, grafted into 7-day-old rats, lost their vesicular follicles and reverted to an infantile condition like that of the host's own ovaries.[299] When gonadotrophin was administered to the host animals, the grafted ovaries responded to it and their follicles grew again, but the host's own ovaries did not, presumably because they had not yet acquired the capacity to do so. It is

possible that the ovary acquires the capacity to respond to gonadotrophin when certain thecal cells begin to secrete oestrogen.[194] The
cycle must start somewhere, but the problem of its initiation remains;
it is discussed more fully in a later section. That oestrogen can affect the
ovary directly, and not only through the feedback mechanisms, has been
shown in several investigations.[360,361]

Oestrus and ovulation

The central event of the ovarian cycle is ovulation, the release of an
oocyte, usually associated with a period of oestrus or 'heat'—the period
of sexual receptivity on the part of the female. This phenomenon plays
so large a part in mammalian reproduction, except in man and the higher
primate species, that it is common to think of the 'oestrous' rather than
the 'ovarian' cycle when one is considering the non-pregnant animal.
It is well to remember, however, that in their natural state animals are
usually either pregnant or anoestrous. The unmated cycle, when the
reproductive organs are 'ticking over', is uncharacteristic. Ovarian
activity is more varied and its rhythm more complex as it responds to
the cycle of the seasons and to a wide variety of changeable and inter-
acting external circumstances.

The hypothalamus and the pituitary gland

Since 1927, when the importance of the pituitary gland in ovarian
function was recognised, its central position in the whole endocrine
system has been emphasized many times ('The master gland of the endocrine system'; 'the conductor of the endocrine orchestra'). The fact that
pituitary function is in turn controlled by the hypothalamus has subsequently been established and the mechanism of this control has, to
some extent, been elucidated. 'It is apparent that the whole pituitary
gland is predominantly subservient to, and partly evolved from, the
hypothalamic portion of the brain'.[373] It is now widely accepted that
the hypothalamus controls anterior pituitary secretion by variable stimulation (or, in the case of prolactin, by inhibition) of a low basic rate of
output. The function of the 'conductor', in the orchestral sense, is pre-
eminently that of an integrator, and it is now clear that this function,
with respect to the endocrine system, belongs to the central nervous
system and especially to the hypothalamic region of the brain.

The structure of the pituitary gland is basically similar throughout the vertebrate phylum (Fig. 3). It develops from the fusion of a ventral diverticulum of the brain (the infundibulum) with a dorsal outgrowth from the pharynx. This outgrowth is originally hollow; it is known as Rathke's pouch, and it is the hypophysis proper. In the ordinary usage of mammalian endocrinology, however, the term 'hypophysis' is used to signify the whole pituitary gland; the posterior lobe of which, derived from the infundibulum of the brain floor, is often referred to as the

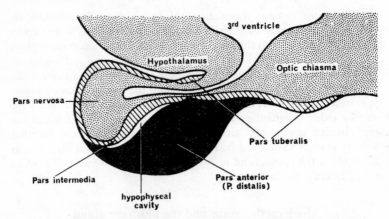

FIG. 3. Diagrammatic sagittal section through the hypothalamic and hypophyseal region of a mammal.

'neuro-hypophysis'. The posterior lobe is also known as the 'pars nervosa', a descriptive term indicating its histology in contra-distinction to the 'adeno-hypophysis', signifying a glandular structure. This is the anterior pituitary, or 'pars anterior' or 'pars distalis'. The anterior pituitary gland (or lobe) is formed from the anterior wall of Rathke's pouch, the posterior wall of which is applied to the pars nervosa and forms the 'pars intermedia'. The upper (proximal) part of Rathke's pouch gives rise to the 'pars tuberalis'. The pars nervosa surrounds a narrow cavity which is continuous with the infundibular recess and hence with the third ventricle. The cavity of Rathke's pouch remains as a narrow 'hypophyseal cleft' between the pars anterior and the pars intermedia. This last-named element, which in lower vertebrates controls colour changes, etc., is reduced and sometimes absent in mammals.

Figure 4 shows the main lines of communication between the hypothalamus and the pituitary gland. Those between the hypothalamus and the pars nervosa (posterior pituitary) are specialized nerve cells, known as 'neurosecretory' cells. All nerve cells are secretory cells in a sense, for they all appear to function by the secretion of 'chemical transmitters' (e.g., acetylcholine) at nerve endings. In the neurosecretory cells, however, this function is enhanced, and the 'transmitter' is passed into

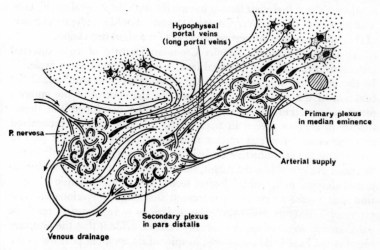

FIG. 4. Diagram of the neuro-humoral relations of the hypothalamus and pituitary gland.

the blood stream as a hormone. This type of endocrine agent is often referred to as a 'neurohumor'. The organ in which the transfer from nerve cell to bloodstream is made is a 'neuro-haemal' organ.

The posterior pituitary secretes substances which are elaborated in nerve cells in the hypothalamus and passed down the axons of these cells directly into the pars nervosa. The communication between the hypothalamus and the anterior pituitary is very different. The anterior pituitary itself elaborates the hormones that it secretes, but it secretes them under the influence of neurohumoral 'messengers' from the hypothalamus. From the diagram it will be seen that the anterior pituitary receives blood not only from a direct arterial supply but also from portal vessels conveying blood from a capillary bed in the median eminence (of the hypothalamus) to a secondary capillary bed in the anterior lobe itself. The significance of these portal vessels lies in the fact that the hypothalamic capillary plexus from which they originate is itself neuro-haemal in function, and the portal vessels therefore carry the neuro-humoral messengers from the hypothalamus direct to the anterior pituitary. A 'portal' vessel is, of course, one that runs between two capillary beds; its function is to convey substances from one organ to another without their first having to pass through the general circulation. The history of the discovery, or perhaps one should say the emergence, of this significance is interesting, and calls for a short digression.

F. H. A. Marshall's famous Croonian Lecture of 1936 directed attention to the effect of external factors on the reproductive cycle.[245] In the course of this study of 'sexual periodicity' he emphasized the function of the central nervous system in the control of the anterior pituitary secretion of gonadotrophins and he developed the theme further in another review in 1942.[246] As it became widely recognized that the central nervous system must play a very important part in controlling the reproductive cycle in all mammals, and not only in the 'induced ovulators' like the rabbit, the possibilities were explored by the use of classical neuro-physiological techniques. Hypothalamic stimulation was found to result in increased discharge of various anterior pituitary hormones, and hypothalamic lesions were shown to result in a decrease in some cases. It gradually became evident that the pituitary gland could have little autonomy, in spite of its apparent position as the 'master gland', and a search was made for a secretomotor innervation whereby the brain might exert a modifying or controlling influence on it. As intensive search failed to reveal any such innervation, the conviction grew that it did not exist. Attention turned again to the possible existence of a humoral control mechanism. This possibility had been rather tentatively suggested already, and it had been shown that ovulation and pregnancy were not affected by bilateral section of the cervical sympathetic trunks in the rabbit.[193] The reasoning behind this experiment was admirably clear. The only known innervation of the anterior pituitary is derived from the carotid plexus, which in turn is supplied almost exclusively from the superior cervical sympathetic ganglion, so that severing the cervical sympathetic trunks would eliminate the most probable path for nervous transmission from the hypothalamus to the anterior pituitary. As the operation clearly did not interfere with ovulation, the pathway must be either along rather obscure cranial nerves to the carotid plexus, or else 'the pathways from the hypothalamus must activate the posterior lobe of the hypophysis, which in turn may exert an influence on the anterior lobe by hormonal transmission'.

There was no reference, in this account, to the hypophyseo-portal vessels, which had already been described as such by Popa and Fielding in 1930.[304] Their description was based on the dissection of human material, and the vessels were referred to as veins 'because they carry the blood away from . . . the pituitary'. Popa and Fielding still assumed the flow to be away from the pituitary when they described their material in greater histological detail some years later. Their diagram of the relation of the portal vessels to the surrounding structures is very clear, and they noted 'heavy glial sheaths of the hypophyseo-portal veins' near their 'ends' in the hypothalamus. It is now known, of course, that these vessels carry blood to, not from, the pituitary, but demonstrating this was not as simple as might be thought. The arrangement has been shown to be substantially similar in all vertebrate species examined; that of the human is well described and illustrated by Daniel.[86]

The possibility that the posterior lobe of the pituitary gland might influence the anterior lobe hormonally, and so pass on a signal from the hypothalamus, was also considered by Harris in 1937.[169] He was able to cause ovulation in the rabbit by hypothalamic stimulation, and he was familiar with the work of Popa and Fielding. His name is prominently associated with the elucidation of this problem and in his book on the *Neural Control of the Pituitary Gland*[170] he gives a most readable account of the anatomical and neurophysiological researches that led to the now accepted view concerning neurohumoral control of the adenohypophysis.

A second set of portal vessels has also been found to run to the anterior pituitary, from a primary capillary bed in the pituitary stalk and the pars nervosa. These therefore provide a possible route for the hormonal control of the anterior lobe by the posterior lobe, as postulated. It is clear, however, that the role of these 'short' portal vessels is subsidiary to that of the 'long' portal vessels running from the hypothalamus to the anterior pituitary. There is good evidence that their primary capillary bed serves as a neuro-haemal organ and that they supply a specific group of cells in the adenohypophysis. Not only is there little mixing of the blood carried to the adenohypophysis by the

C

58   THE OVARIAN CYCLE OF MAMMALS

long portal vessels with that carried by the short portal vessels, but there is evidence that among the long portal vessels, individual ones supply well demarcated areas of the gland, so that blood leaving a particular group of cells in the hypothalamus may be conveyed to a specific group of cells in the anterior pituitary. It has been stated that the long and the short portal vessels are the sole sources of blood of the pars distalis in the rat, sheep, goat, man and monkey,[233] but this is not true of the rabbit gland.[86]

The anterior pituitary hormones include growth (or somatotrophic) hormone (GH or STH) and prolactin, as well as adrenocorticotrophin (ACTH), thyroid-stimulating hormone (TSH) and the gonadotrophins (FSH and LH). All except prolactin are released under stimulation by the neurohumors which are carried by the hypophysial portal vessels and are therefore identified as 'releasing factors'. The release of prolactin is also under hypothalamic control: in this case the controlling factor is not stimulatory but inhibitory and is therefore known as prolactin-inhibiting factor (PIF). This fact may be related to the complex synergism that appears to exist between growth hormone and prolactin. The difference between the mode of control of FSH and LH and that of prolactin is exploited in the experimental method of comparing ovarian response after hypophysectomy and after hypophysial stalk section (p. 99). The former deprives the organism of all pituitary hormones, whereas the latter reduces the release of gonadotrophins but not that of prolactin.

The nature of the effect exerted by the hypothalamic releasing factors, at cell level, has not yet been established. They may act solely by altering the condition of the cell membranes and thus controlling cell permeability to different ions, or by directly affecting the release of stored product. On the other hand, they may actually take part in the biosynthesis of the pituitary hormones. No direct evidence of such participation, however, such as the incorporation of labelled amino acids into a specific hormone, has so far been produced. A comprehensive review of the current state of information about hypothalamic releasing and inhibitory factors is provided by McCann and Porter.[234]

It has been shown that the releasing factors are highly specific, and it is probable that their respective modes of action may be widely different. For example, the release of FSH under the stimulus of FSH-RF is blocked by substances that inhibit protein synthesis (puromycin, actinomycin) but this is not true of LH. It is therefore argued that FSH release somehow involves the synthesis of protein *de novo*, whereas LH release does not.[232] The releasing factors normally operate 'locally', being transported directly from their sites of origin, or perhaps from storage depots in the median eminence, to the target (pituitary) tissue. But they are truly humoral, or blood-borne, and some of them have been detected in the peripheral blood of hypophysectomized animals. The pituitary gland retains some secretory activity after auto-transplantation to a distant site, and this residual activity appears to be in response to the attenuated supply of releasing factors that reach it in circulating blood. For example, anterior pituitary gland tissue trans-planted to a position under the skin, in a rat, has been shown to secrete increased amounts of gonadotrophin (FSH) when the rat was kept in conditions of continuous illumination. This stimulus presumably oper-ates through the hypothalamus; it has the same effect on the normal pituitary. In the rats bearing pituitary autotransplants the gonadotrophin releasing factor(s) must have reached the pituitary tissue by way of the systemic circulation. There is confirmatory evidence that this is so, since the activity of autotransplanted hypophyseal tissue is abolished by destruction of the median eminence.

None of the gonadotrophin-releasing factors has so far been 'characterized' (in the biochemist's jargon) as completely as thyro-trophin-releasing factor. This substance, which has a molecular weight below 4,000, has been obtained in a 'highly purified' form, but it has not yet been chemically defined. This is partly, at least, because it is available in such minute amounts. It is effective in nanogram quantities, even as a single injection but, in one assault on the problem the extrac-tion of several million specimens of sheep hypothalamus yielded only about 1 mg of the highly purified substance. Thyroxine exerts a restraint on the secretion of TSH by a negative feedback mechanism affecting the output of TRF from the hypothalamus. This is similar to the negative feedback control of gonadotrophin secretion and ACTH secretion, but whereas the steroid hormones of the gonads and adrenal cortex appear to affect the hypothalamus directly (Fig. 5a), thyroxine apparently acts on the pituitary to induce the synthesis of a polypeptide which inhibits the secretion, or perhaps the action, of TRF (Fig. 5b).

The running activity (measured in a wheel cage) that is typical of oestrous behaviour in the rat has been used to detect the effect of oestrogen implanted into various regions of the hypothalamus in ovari-ectomized rats.[60] Running behaviour was induced by such implants, and those in certain locations were more effective than others, but the response was not so well localized that a particular response 'centre' could be identified. The synthetic steroids used in contraception have also been shown to inhibit gonadotrophin secretion by their effect on the hypothalamus, in much the same way that the gonadal and adreno-cortical steroids do.

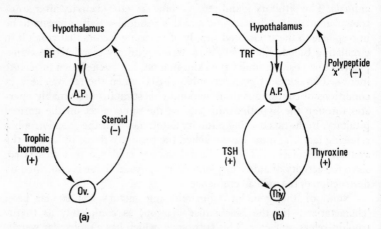

Fig. 5. Direct (*a*) and indirect (*b*) feedback pathways from target organs to the hypophysis and hypothalamus.

A further refinement in the hormonal mechanism controlling ovarian function has been demonstrated within the last few years by experiments showing that the pituitary gonadotrophins may themselves exert an inhibiting influence (negative feedback) on their own hypothalamic releasing factors.[46] Implants of LH or FSH, placed in the median eminence of the hypothalamus, lowered the level of the respective hormone in the pituitary tissue, and it was concluded that such 'short feedback' mechanisms play a part in normal physiology.[88] The postulated relation between the 'long' and 'short' feedback loops is shown in Fig. 6.

The establishment of cyclic gonadotrophin secretion

In 1935, Pfeiffer described some experiments that pointed to the early establishment of a cyclic (female) or non-cyclic (male) rate of gonadotrophin secretion by the pituitary gland.[297] He found that when an ovary was transplanted to a male rat that had been castrated at birth, it (the ovary) underwent cyclic changes. When a testis was transplanted to a female rat very soon after birth, it prevented the cyclic activity of the recipient's ovaries, or of ovaries transplanted into her in adult life

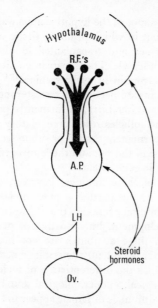

Fig. 6. Long and short hypothalamic feedback loops.

from an untreated female. The significance of these results, in terms of the differentiation of male and female sexual function, was apparently not recognized by other endocrinologists until about 20 years later, when attention was focussed on hypothalamic, rather than pituitary, function. This is the more surprising since Pfeiffer clearly stated that the experiments were undertaken to show whether the functional difference between male and female hypophyses is 'a primary difference . . . determined by the sex genes, or a secondary difference which is determined by the gonad function and is, therefore, dependent on whether an ovary or a testis is present'. Pfeiffer concluded 'that the hypophysis is permanently altered' and 'that the sex type of the hypophysis is secondary, depending upon the presence of differentiated sex glands'. He re-stated this conclusion rather more bluntly in a subsequent publication: 'The hypophysis in the rat at birth is bipotential and capable of being differentiated as either male or female, depending upon whether an ovary or a testis is present.[298]

The central importance of the hypothalamus was emphasized by the result of an experiment in which pituitary tissue was taken from very young rats of either sex and transplanted in place of the pituitary removed from an adult female. The transplanted pituitary tissue thenceforth secreted gonadotrophin in cyclic fashion.[171] Shortly afterwards, it was found that a single injection of testosterone, given to a female mouse on the fifth day after birth, permanently prevents ovulation.[27] The ovaries grow, and follicles reach preovulatory size, but the animal remains in a state of permanent oestrus or suboestrus. It was later found that the same was true of the rat, and the 'androgenized female rat' has been extensively used in experiment.

The significance of these findings, however, extends beyond the laboratory usefulness of this particular phenomenon. It is not only the definitive pattern of pituitary function that is imprinted during the brief critical stage of development, but also that of male or female sexual behaviour. Although these attributes are not manifested until after puberty, their character is evidently determined at about the same time, and in the same way, as the differentiation of male and female genital tracts. The morphological, endocrinological and behavioural patterns characteristic of the adult male or female appear to be selected according to whether or not small amounts of androgen (testosterone) are circulating during a critical few days in late foetal or early postnatal life. This phenomenon of 'imprinting' will be discussed further in the succeeding section, dealing with hypothalamic function in relation to sexual behaviour.

The hypothalamus and sexual behaviour

We have seen that mammals respond in characteristic ways to environmental changes perceived through their sense organs, that the hypothalamus governs the secretory activity of the pituitary, and that a balance is maintained between hormonal stimulus and response within the endocrine system. All the stimuli from the external environment that are perceived by the sense organs must be channelled through the hypothalamus, either directly or by way of the cerebral cortex. The ovarian hormones not only affect the uterus and other sexual organs, and exert through the hypothalamus a feedback effect on their own secretion rate. They also stimulate behavioural responses, and the hypothalamus is again involved. Fig. 7 illustrates the pathways involved in simple outline. The connection indicated between behavioural response

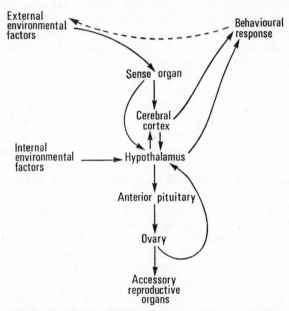

FIG. 7. Diagram to illustrate the central position of the hypothalamus in the regulation of behaviour in response to external stimuli, as it affects the reproductive system.

and environmental factors is included because the animal exercises a degree of control over its environment. The behavioural response may, for example, be such as to extinguish the environmental stimulus, or remove the animal from its influence.

Having examined the pathways of communication in the successive relays within the organism, as far as the hypothalamus, we should, logically, proceed to consider the ways in which stimuli from the sense organs are conveyed to the hypothalamus, the mechanism by which a response is invoked in the hypothalamus, and the means by which it governs the animal's behaviour. Here we encroach upon a very difficult area of investigation, but it is clear that the hypothalamus fulfils both an administrative and an executive role.

Most of the stimuli that reach the hypothalamus from the sense organs are transmitted to it through a variety of 'centres' in the cortex or thalamus, and the hypothalamus may influence behaviour by stimuli that are similarly transmitted through the 'higher' centres. On the other hand, experiments involving ablation of the cortex have shown that many of the behavioural responses concerned with reproduction are controlled by hormones acting directly on the hypothalamus, and cortical structures are not necessarily involved in either the afferent or the efferent pathway.

That oestrogen causes heat is, of course, implied in the name of the hormone, and ovariectomized animals do not exhibit oestrous behaviour unless oestrogen is administered (in some species progesterone is required as well). That oestrogen can act directly on the hypothalamus has been shown by experiments in which a minute amount of it was fused to the tip of a needle which was then implanted stereotaxically in a pre-determined position in the brain. The stereotaxic technique employs a mechanical device to carry a probe or similar instrument to an exact position relative to a rigid frame. The head of the experimental animal is held in position within the frame and, as the skull is rigid, the position of any organ or region within it can be related to the geometry of the frame, the traverse of the probe can be recorded, and any manoeuvre can be accurately repeated.

The first investigations of this kind into sexual behaviour were carried out on spayed cats. The amount of oestrogen needed to produce oestrus when implanted into the mammillary body (in the posterior part of the hypothalamus) was very much smaller than the amount required in order to produce the same effect by systemic administration. These small amounts had no discernible effect on the genital tract. In subsequent work, the technique has been widely applied to determine the site of action of a number of substances in various species. The reports indicate differences between species, but it is not possible to say, for instance, that oestrous behaviour is controlled by a particular hypothalamic body or nucleus in a particular species. The mechanism is complicated, and many modifying influences may be called into play.[48]

THE OVARIAN CYCLE   65

Martini and his colleagues in the University of Milan have reviewed some aspects of the integrative function of the hypothalamus, with particular reference to their own work on the location of the feedback receptors and the cells synthesizing the releasing factors.[249, 250] Their experiments indicated that the median eminence is the region particularly concerned with receiving 'feedback messages' and that the releasing factors are synthesized elsewhere. If this is so, there must exist nervous connections between the median eminence and the peripheral regions of RF synthesis. They suggested that such messages may well be passed, first, to intermediary centres where the 'feedback' information can be integrated with that received from external sources, before a message is finally directed to the RF-synthesizing cells.

Having determined that oestrogen can have a specific effect in the hypothalamus, the question arises whether this is the way in which naturally occurring oestrogen, or systemically administered hormone, elicits oestrous behaviour. Does it actually reach the hypothalamus in diluted amounts, and so function like the minute hormonal implants? There is some evidence that it may, because hypothalamic neurons have been shown to take up oestrogen that has been administered systemically. This was demonstrated by the administration of isotopically labelled oestrogen; its presence, and its exact position, in the brain tissue was subsequently determined by autoradiography.[254] In this technique, sections (slices) of the tissue under examination are placed on photographic emulsion, which records the radioactivity due to any 'labelled' substance that has been incorporated into the tissue. It can be traced in individual cells, and even located in intracellular organelles.

The endocrinology of oestrus has been investigated in more detail than that of other behavioural phenomena in either sex. This is partly because oestrus is readily distinguished in the course of experiment, and partly because it is more completely governed by the sex hormones than is, for instance, maternal behaviour. The latter has nevertheless been studied in rats and rabbits, and it has been shown that nest-building by female rabbits is under endocrine control. It is abolished when the ovaries are removed, but it can be induced in spayed animals by appropriate hormonal treatment. This was demonstrated in experiments in which the hormones were administered systemically, but it seems probable that their mode of action in this function, as in the induction of oestrus, is through the hypothalamus.

There appears to be a sex-specific pattern of behavioural response, imprinted at an early stage of development. We have already seen that the prepuberal differentiation of male or female genitalia, and of cyclic or non-cyclic secretion of gonadotrophin after puberty, are both determined during a critical stage of foetal or early postnatal life. We have seen, furthermore, that in both cases the organisation of male or female characteristics depends on the presence or absence of circulating androgen. It has more recently been shown that a parallel situation exists with respect to the differentiation sometimes referred to as the 'organization' of behavioural response. Here, too, the sex-specific pattern, determined in early life, appears to remain unactivated throughout the intervening period until puberty. The positive effects again appear to be towards the establishment of maleness, so that female characteristics develop in default of male hormone. It is almost as though the basic mammal were a female, the male being a modification depending on the presence of the 'Y' chromosome. It is interesting that the reverse is true of birds, where the male is the homogametic sex and the development of the ovary is hormonally imposed, that of the testis being anhormonal. One is therefore tempted to think it significant that the hormonally imposed sex should be the heterogametic one, but any such speculation has to be hedged about with reservation, for the evolution of the sex chromosomes and their relation to the sex hormones is far from fully understood. It is clear that the gonadal hormones have a very special role in mammals; it is conceivable that their function in the differentiation of the sexes early in development preceded their adult function phylogenetically as well as cytogenetically.

Experimental evidence of the 'imprinting' of sex-specific behaviour was provided by experiments using an 'anti-androgen', a drug that inhibits, or antagonizes, the action of testosterone. The type of mating behaviour displayed by normal adult female rats was suppressed by the administration of such a drug. When the anti-androgen was administered to pregnant rats, the genetically male offspring were found to exhibit 'female' mating behaviour. If subsequent investigation shows them to be totally incapable of male behaviour, this will provide very strong evidence that a male behaviour pattern is imprinted during foetal life and that a female behaviour pattern is inherent in the organism, regardless of its genetic sex. Earlier experiments showed that the administration of androgen to female rats during the first few days after birth, made them incapable of 'female' behaviour after puberty. The critical period is not so well defined as that governing the differentiation of 'male' or 'female' gonadotrophin secretion, and larger amounts of androgen were required to overcome the 'female' pattern. Male rats castrated at birth retain some capacity for 'male' behaviour under the influence of testosterone administered in adult life. This 'residual maleness' may represent an inherent capacity, but if it is found to be lacking in males subjected to anti-androgens *in utero*, as described above, it may be attributed to minute amounts of androgen circulating before birth.

All through the foregoing descriptions of hypothalamo-hypophyseal relations, emphasis has been placed on the evidence which suggests that the steroidal hormones affect the pituitary through the hypothalamus. All the experimental data, including the pioneer work of Pfeiffer, appear to be consonant with the idea that this is so. Several authorities, however, have warned against accepting the evidence as finally conclusive, strong though it is. It is difficult to exclude the possibility of 'leakage' or transport of hormones to the hypophysis, and there is direct evidence that oestradiol implanted in the median eminence may be transported to the pituitary gland. This is not to say that evidence from experiments based on hypothalamic implants should be discounted. It is, rather, an indication of the care required in the interpretation of experimental data, especially in such a complex and sensitive area.

The ovarian cycle of the laboratory rat

When one reads of an experiment having been performed on a 'rat', the animal referred to is a laboratory-bred specimen derived from *Rattus norvegicus*, the heavier and coarser-haired of the two common species (the other is *R. rattus*, the black rat, or ship rat). The laboratory mouse is similarly derived from an albino strain. The habits and breeding of these species are described in detail in several well known volumes, references to which will be found in the valuable UFAW Handbook (on the care and management of laboratory animals). The white rat was developed as a laboratory animal about 50 years ago in the Wistar Institute in Philadelphia.[104]

The 'normal' cycle

The female rat, when maintained in good condition but isolated from males, ovulates at intervals of 4 to 5 days. Ovulation occurs during oestrus, and oestrus can readily be detected in several ways. The method most commonly used is the 'vaginal smear'. A spatula, or a smooth rod, or a cotton-wool swab, is inserted into the vagina to collect a sample of the cells of the vaginal wall, and those lying free in the lumen; this is 'smeared' onto a microscope slide. The 'smear' is air-dried and may be stained and cleared, and mounted under a cover glass. If all that is required is to follow the cycle daily, simple staining and immediate examination will suffice. While the animal is in oestrus the smear will consist of cornified cells sloughed off from the vaginal wall.

A. S. Parkes recalls that during G. F. Marrian's classic work on hormone extraction from human urine, 'Inadequate resources even led, on one occasion, to a spatula being used successively for chemical work and vaginal smearing. This was in the days before the extraordinary efficiency of local application of oestrogen was known and the resulting epidemic of vaginal cornification among the spayed mice was very puzzling.'[283] When normal intact rats or mice are used, it is all too easy to induce pseudopregnancy (see below) by inadvertently stimulating the cervix while taking a vaginal smear.

Long and Evans in their classic description of the oestrous cycle of the rat,[228] recognised four stages: oestrus, metoestrus, di-oestrus and pro-oestrus. More recently the successive stages of the cycle have been studied with particular reference to the vaginal smear, 'in view of the need to time various operative procedures accurately in relation to the phases of the oestrous cycle.'[240] Six stages in the cycle may be recognized in terms of the vaginal cytology, as follows:

1. Early oestrus (about 18 hr); a thick smear, mainly of basophilic nucleate epithelial cells.

2. Oestrus (about 25 hr); epithelial cells have lost their nuclei and become cornified. Towards end of oestrus, smear becomes 'cheesy'. Cornified cells are acidophilic.

3. Late oestrus or metoestrus (about 5 hr); many cornified cells, but also some large basophilic ('Shorr') cells and some small basophilic cells.

4. Early di-oestrus (about 24 hr); a thick smear, mainly of leucocytes.

5. Di-oestrus (about 28 hr); a thin smear, mainly leucocytes.

6. Late di-oestrus (about 7 hr); leucocytes and some clearly nucleate basophilic epithelial cells.

It will be noticed that 'dioestrus' (including 'early' and 'late' stages) is characterized by the presence of leucocytes in the uterine lumen. The dioestrous part of the cycle is more variable than pro-oestrus and oestrus. Ovulation occurs during late oestrus ('metoestrus').

The periodicity of the cycle in the unmated rat or mouse is short because there is nothing to delay the maturation of a second crop of follicles after the rupture of the first crop. When the follicles rupture, the level of circulating oestrogen is suddenly reduced. Corpora lutea are formed from the ruptured follicles (as always) but in the absence of mating or a similar stimulus, in these two species and in few others, the corpora lutea are 'inactive'—i.e., they secrete only a very small amount of progesterone. Both oestrogen and progesterone levels are therefore low, and new follicles are ovulated as soon as they have 'ripened' sufficiently.

This short cycle has been intensively studied in the laboratory rat. It is relatively uncommon among wild rats, which are likely to mate whenever they enter oestrus, but in laboratory conditions it has come to be known as the 'normal' cycle, and it is so labelled in Fig. 8.

The luteal phase: pseudopregnancy

The simplest variation from the 'normal' cycle, in the rat or mouse, is one in which the corpora lutea are active so that a 'luteal' phase is interposed between successive periods of heat. This occurs after a sterile mating, or after mechanical or electrical stimulation of the cervix uteri near the time of ovulation. In the laboratory this condition is readily brought about by housing the female with a vasectomized male, so that mating will occur without fertilization and conception. The luteal phase induced by any such means is known as a 'pseudopregnancy', and in the rat it can only be induced by a neural stimulus transmitted through the cervix, as described. In most species, on the other hand, a similar luteal phase, with active corpora lutea, is an invariable component of the cycle, whether the animal is mated or not. Because the luteal phase in such a cycle is so like that of pseudopregnancy in the rat, one sometimes sees it referred to as a 'pseudopregnant cycle' although no cervical stimulation or sterile mating is involved. There is one important difference, although it is not often emphasized: pseudopregnancy in the rat is more variable in duration than, say, the 17-day cycle of the guinea-pig or the 21-day cycle of the pig. Both, however, are alike in their most significant feature, for in both of them the uterus undergoes the same changes (as far as is known at present) as it would undergo if fertile mating had occurred. The early stages of pseudopregnancy, or the luteal phase, are indistinguishable from the first days of pregnancy.

'Progestation'

The changes that occur, particularly in the uterus, are referred to as 'progestational' changes, and in recent years the term 'progestation' has come into use to signify the early stages of pregnancy or pseudopreg-

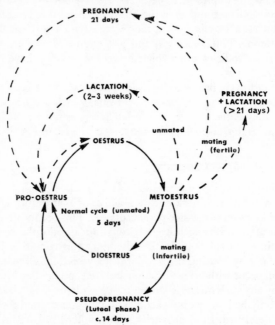

FIG. 8. The ovarian cycle in the laboratory rat.

Oestrus (and ovulation) recurs every 4 to 5 days in the unmated animal. Infertile mating or other cervical stimulation extends the interval to about 14 days. Fertile mating leads to pregnancy (21 days) and if the rat becomes pregnant again at the immediately post-partum oestrus the gestation period may be lengthened by the concurrent lactation. Lactation unaccompanied by pregnancy may postpone oestrus for as long as pregnancy without lactation.

nancy, particularly the uterine changes preparatory to implantation. It is used by some writers not only to indicate the processes within the uterus, but as a convenient term for the pre-implantation phase of pregnancy, the prefix 'pro' having the meaning of 'preceding'. There is some danger of confusion here, for if 'progestation' is the part of pregnancy that precedes implantation, the time from implantation to parturition should be 'gestation'. One has often wished that this word could legitimately be used in this restricted sense, but unfortunately it is clear that 'gestation' is well established as a synonym for 'pregnancy', and means the whole of the period from fertilization to parturition. Nevertheless, 'progestation' appears to have come to stay, and one must hope that the advantages of its convenience outweigh the disadvantages of its ambiguity.

Pregnancy

In the rat, and in the mouse and guinea-pig, a 'vaginal plug' is formed after mating by the solidification of the secretions of the male accessory sex glands. Coitus usually occurs at night and the plug may be found *in situ* or, more frequently, on the cage floor the following morning.

The corpora lutea of pregnancy or pseudopregnancy are not, at first, different in size or in histological appearance from those of the unmated cycle, even in the rat. It appears indeed, that the female mammal does not, so to speak, 'realise' that she is pregnant until gestation is well under way. This is particularly true of animals in which implantation of the embryo is superficial and does not involve early nidation and the active proliferation of endometrial (uterine) cells. We shall return to the question of 'how does the mother know she is pregnant?' at a later stage (p. 109). The uterine changes in the early part of the luteal phase in the unmated guinea-pig, and in the pseudopregnant rat, are such that the uterus responds to trauma (e.g., puncture of the wall, or filling the lumen with oil) by the production of 'deciduomata'. These (the singular is 'deciduoma') are swellings in the uterine mucosa almost exactly like those that are formed in pregnancy when the blastocyst settles and its trophoblast attacks the uterine epithelium, so that the embryo appears to burrow into a cushion of uterine ('deciduomal') tissue. The term 'decidua' of course implies tissue that is shed. It is, in fact, the maternal contribution to the foeto-maternal placenta; the 'shedding' occurs when the placenta is voided as the afterbirth, following the expulsion of the foetus at parturition. Deciduomata experimentally induced during pseudopregnancy are mainly resorbed in the uterine wall, but there is some breakdown and loss of tissue and blood. It is doubtful whether this bears any significant relation to menstruation in the primates (see p. 75). No such breakdown of tissue occurs in pseudopregnancy if deciduomata are not provoked.

Pregnancy constitutes a major variant of the ovarian cycle in the generality of mammals, including the rat. If embryos implant, and pregnancy proceeds normally, the luteal phase is prolonged beyond its duration in pseudopregnancy. Pregnancy lasts 21 days in the rat, and there is a dramatic increase in the size of the corpora lutea about the 13th–14th day (Fig. 9). They nearly double in volume at this stage, and in spite of the vast amount of investigation that has centred on the rat's cycle, the stimulus and the 'reason' for this luteal growth remain unknown.

Plate V shows a series of photographs of sections through some of the ovaries from which the data shown in Fig. 9 were derived. The animal killed $1\frac{1}{2}$ days after mating (a) has surviving corpora lutea from a previous cycle, and one of the new corpora lutea is shown, still hollow. A little later (b) the older corpora lutea have noticeably regressed and the new ones, now solid, are growing. They are easily distinguishable at higher magnification. Eleven days after mating, (c) the old corpora lutea have regressed almost completely, and more follicles are growing. Fourteen days after mating, (d) the corpora lutea of pregnancy have enlarged; the follicles remain in the 'resting' condition. Just before parturition (e) the corpora lutea have begun to regress; follicles have ripened and are about to ovulate.

The post-partum oestrus

The diagram in Fig. 8 shows still further ovarian variations. The corpora lutea of pregnancy regress immediately before parturition and another crop of follicles matures. These follicles rupture within 24 hours of parturition, and their ripening is accompanied by oestrus. This is the 'post-partum' oestrus characteristic of rats, mice, guinea-pigs and some other animals. The first oestrus after parturition in other animals, like the pig, sheep or horse, is sometimes referred to as the post-partum oestrus, although it occurs days or weeks afterwards. The expression was originally used to denote oestrus within a very short time after parturition, for this reason one may sometimes read of the common domestic animals that 'there is no post-partum oestrus'.

If the female rat mates at the post-partum oestrus she may, and often does, become pregnant while lactating. If she is suckling more than a few young, pregnancy lasts longer than its 'normal' 21 days, because implantation is delayed. This phenomenon of 'delayed implantation', however, is a normal feature of pregnancy in some species, the

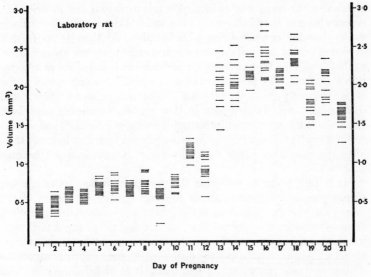

FIG. 9. Growth of the corpora lutea of pregnancy in the rat.

Measurements of the individual corpora lutea in the ovaries of rats killed at a known stage of pregnancy, measured from the day of mating (day 0) when a vaginal plug was found. Each column represents one rat, each line one corpus luteum, so the variation within each animal is shown, and the average luteal volume for each animal can be assessed.

term being used to denote that, in these species, the blastocyst lies free in the uterine lumen for some considerable time, with little or no growth or cell division, before it implants. In a rat that is suckling a large litter of ten or more young, the blastocysts may remain free in the uterus for seven or eight days, instead of implanting after about a day. The delay is less when fewer young are being suckled, and it is quickly terminated (by implantation) if the litter is removed. In some of the species in which implantation is invariably 'delayed', the blastocyst may remain unattached for months. The endocrinological implications of such delay, and the factors involved in terminating it, have been the object of much investigation in recent years, and will be discussed in greater detail later (p. 118).

If the rat does not mate at the post-partum oestrus, or if the mating is infertile, the corpora lutea formed at this ovulation, and described as corpora lutea of lactation, persist for a variable time, some two or three weeks, roughly corresponding with lactation. As there is no further ovulation or oestrus until they regress and cease to secrete progesterone, so that fresh follicles can mature, the condition is described as a 'lactation anoestrus'. The persistence of the corpora lutea of lactation in the rat is apparently due to the output of prolactin associated with lactation, since prolactin is luteotrophic in this species, as already described. Prolactin is apparently not luteotrophic in the guinea-pig*, and oestrus recurs about 17 days after the immediately post-partum heat—that is, after the normal oestrous interval—if the animal does not become pregnant.

It is interesting to note that the post-partum heat of rats, mice, guinea-pigs and rabbits precedes the post-partum involution of the uterus and vagina. Parturition in the guinea-pig involves the relaxation, almost the 'melting' of the pubic symphysis, and the young at birth each weigh about a quarter of the weight of a young adult, so the uterus, cervix and vagina are enormously distended. Oestrus nevertheless occurs within an hour or two after parturition. It is of brief duration, and copulation is accomplished rapidly, but both the ovulation rate and the conception rate are high at this oestrus, so that it is a highly fertile heat period in this species.

Variants of the eutherian reproductive cycle

It will already be apparent that the features of the ovarian cycle that are common to the generality of mammals are limited to its basic elements. The variations, however, fall within well defined major groups, which may be summarized as follows:

1. Species with a menstrual cycle.

2. Species with an oestrous cycle:
    (a) Ovulation dependent on mating,
    (b) Spontaneous ovulation with 'inactive' corpora lutea,
    (c) Spontaneous ovulation with functional corpora lutea.

*But see p. 35 (footnote).

Not surprisingly, since the hypothalamic-pituitary-ovarian control system is similar in all of them, the dividing lines between these groups are not absolute. True menstruation and the absence of a 'heat' period, however, is solely characteristic of the higher primates. In the lower primates and in all other eutherian species, the climax of the follicular phase is oestrus, or heat, which bears a constant temporal relation to ovulation in each species. Ovulation occurs just before or just after, or during, oestrus. In the higher primates mating is not confined to a particular phase of the cycle, and these species therefore do not have an 'oestrous' cycle. The basic ovarian cycle follows the same pattern, but whereas the end of the follicular phase culminates in oestrus in all the other species, ovulation is not accompanied by any overt or external sign in the higher primates. In them, the periodicity of the cycle is manifested in the phenomenon of menstruation, which occurs at the end of the luteal phase—diametrically opposite, as it were, to the time of ovulation. The comparison will be carried a little further when the endocrinology of the oestrous cycle is discussed.

The primates

The distinction between species with an oestrous cycle and those with a menstrual cycle, in relation to primate taxonomy, has been set out in the following terms: 'There is a sharp distinction between strepsirhines and haplorhines. The former have a definite oestrous cycle with well defined periods of oestrus with a closed vagina . . . Such vaginal closure does not occur in haplorhines . . . the haplorhine reproduction cycle terminates in menstruation and not by the regression of the prepared endometrium as in all other mammals.[58] The Strepsirhini include the Lorisoidea and Lemuroidea; the grade Haplorhini comprises the Tarsoidea and Pithecoidea.[190]

Although the haplorhine primates are especially characterized by menstruation, the breakdown and renewal of the uterine epithelium is not accompanied by an overt blood loss in all of them. In the New World monkeys, in particular, it is represented only by the shedding of a few red blood cells, not detectable macroscopically. In these species, too, it is clear that copulation occurs more frequently near the time when ovulation is to be expected[136] and it occurs during a more restricted period in New World than in Old World monkeys. Among the strepsirhines, although the bush-baby *Galago senegalensis* does not menstruate, its endometrium has a dual arterial supply very like that found in menstruating species.[58] Rao[309] described a 'microscopic menstruation' in the slender loris, very like that of the New World haplorhines, but later investigators have not confirmed this. Ioannou has reported a restricted mating period and no menstruation in two female pottos (Lemuridae) kept in captivity,[205] and Manley has described the oestrous cycle in five lorisoid species and concluded that 'the group is fundamentally polyoestrous throughout the year.'[242]

Tupaiidae

Uterine bleeding and desquamation of the uterine lining has been described in several species of tree shrews (Tupaiidae) and it has been suggested that it is a true menstruation. This family has both primate and insectivore characters; present-day taxonomists tend to include its members in the Strepsirhini or Prosimii, as relatives of the lemurs. A colony of tree shrews, collected in North Borneo, was established in captivity at the University of Missouri in 1962.[77] The colony has since been augmented by breeding and by further collection, and both breeding behaviour and reproductive physiology have been studied in a number of species.[366] These animals exhibited a well marked oestrous cycle, with intervals of either 10 to 12 days or 20 to 22 days in *Tupaia longipes* and a similar pattern in other species. It was thought that the short cycles had no active luteal phase, whereas ovulation, corpus luteum formation and progesterone secretion were involved in the longer ('pseudopregnant') cycles. The end of such a pseudopregnancy was marked by the recurrence of oestrus and this was soon followed by 'menstruation'. This involved endometrial desquamation and bleeding, and resembled true primate menstruation in being caused by progesterone withdrawal. It was described as being 'very similar to that occurring in South American monkeys'. To equate it with the menstrual bleeding of monkeys and apes, however, one has to suppose that in *Tupaia* the follicular phase, which in man lasts as long as the luteal phase, is somehow telescoped into the latter so that ovulation occurs before the decidual (endometrial) desquamation. This comparison is made diagrammatically in Fig. 10. It was found that oestrus also occurred immediately after parturition, or even before it, the females attracting attention from the males often to the detriment of the young.

In the shorter oestrous cycles that tree shrews sometimes experienced

in captivity, ovulation either did not occur or did not lead to the formation of an active corpus luteum in the absence of a coital stimulus. It is
probable that neither the 'inactive' cycle nor the 'pseudopregnant' cycle
occurs in the wild state. Their normal reproductive pattern is probably
one of repeated pregnancies during the breeding season. Such a difference between free and captive animals is very general, and has been

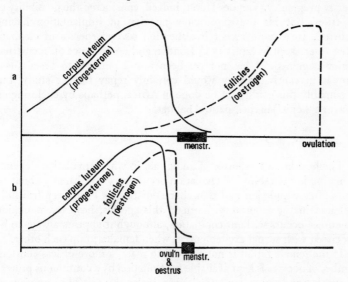

Fig. 10. Diagrammatic comparison of (a) haplorhine and (b)
tupaiid cycles.

expressed in these terms: 'The frequent occurrence of non-pregnant
and pseudopregnant cycles in any species should be considered an
artifact of captivity. Such cycles result from maintaining animals in
abnormal conditions such as isolation or crowded cages . . . In caged
domestic rabbits, estrus has persisted for long periods and females have
copulated during pregnancy and pseudopregnancy. When maintained
under more natural conditions in larger enclosures receptivity was
confined to regular discrete periods'.[366] The effects of cage population
density and the like will be discussed more fully in relation to the breeding season and the effects of external factors and 'social' influences
(p. 160).

'Menstruation' has also been attributed, probably mistakenly, to another family of insectivores, the Macroscelididae or elephant shrews.[377, 379] In them, as in the tree shrews, a decidual cushion or 'polyp' is formed in the endometrium at the pre-determined site of implantation. It seems probable that the shedding of such polyps, in the specimens described by Van der Horst, followed the loss of the embryos in early pregnancy. Van der Horst, indeed, considered the possibility of 'abortion' but his drawings show no sign of implantation having occurred. In a more recent investigation several species of elephant shrews were kept in captivity in London and no evidence of 'menstruation' was gained. It has not yet been found possible to breed these animals in captivity; they would certainly repay further study, but presumably this could best be done in Africa, perhaps by enclosing an area (or a 'kopje') in their natural habitat.

'Induced' or 'reflex' ovulation

The best known example of an animal in which ovulation depends on mating is the rabbit.[180] Ovulation occurs very consistently about 10 hours after copulation and this characteristic has been very fruitfully exploited in the laboratory. Even in this species, however, ovulation sometimes occurs without copulation, although it appears always to be associated with sexual excitement, as when females jump each other.[115]

If the female rabbit is not allowed to mate, a more or less constant number of oocytes is kept available for ovulation by a continuous process of ripening and regression (atresia) of follicles.[189, 359] The same is probably true of other species in which ovulation depends on a coital stimulus—the animals sometimes rather loosely referred to as 'induced ovulators'. Among them are the ferret,[244] mink,[166] ground squirrel,[137] short-tailed shrew[288] and, probably, the European weasel,[93] the American mole *Scalopus aquaticus*,[76] the American opossum[247] and the domestic cat. Oddly, in view of its ubiquity, there is still some doubt about the last-named species, but 'It is generally considered that ovulation in the cat is the result of a neuro-humoral process'.[341]

'Induced ovulation' may be more widespread than has generally been thought. It now seems certain, for example, that it is characteristic of the common field vole *Microtus agrestis*, at least in captivity. It was formerly stated that this vole was polyoestrous and that its cycle closely resembled that of the mouse and rat. This conclusion was based on the fact that wild voles were found at the beginning of the season with more than one set of corpora lutea in the ovaries. Assuming that infertile mating would lead to pseudopregnancy, in which case successive sets of corpora lutea would not accumulate in the same way, it was concluded that they were formed at successive periods of oestrus when no mating occurred.[42] A similar observation was made with respect to the bank vole,[43] but this species does not seem to have been re-examined as the field vole has been, using a well established laboratory population. In such a population of field voles, no virgin females were found to have corpora lutea, but mating with either an intact or a vasectomized male was invariably followed by ovulation.[44] There was no immediately post-partum oestrus, but lactating voles were readily mated on any day of lactation from the second onwards. Similar observations were made on voles kept in a large outdoor enclosure under conditions that appeared to approximate closely to their natural habitat.

It seems impossible to reconcile these observations with the earlier account of multiple sets of corpora lutea in wild voles, unless such cycles occur only at the onset of the breeding season or unless 'domestication' has eliminated this type of periodicity in the animals' reproductive behaviour. 'Induced ovulation' has recently been described in a laboratory stock of another vole, *Microtus ochrogaster*, derived from a breeding stock captured in Missouri in 1959, and in two other American species of the same genus observed in the wild. This type of cycle was also attributed to *M. guentheri* in the Middle East by Bodenheimer in 1949; he put forward the then surprising suggestion that the reproductive cycle was affected, and breeding initiated, by specific dietary factors.[32] In the habitats from which these animals were taken, spring is a sudden phenomenon, and conditions change almost overnight from desert to lushness. In the light of further work on nutrition and breeding (see p. 169) a correlation between this change and the onset of breeding is less astonishing, but the suggestion that the vegetation included a hormone-like substance, acting directly upon the pituitary-ovarian axis, has not been substantiated.

Another candidate for inclusion among the 'induced ovulators' is the camel, an interesting domesticated animal about whose reproductive cycle relatively little is known with certainty. An Egyptian research team, however, has recently issued a report of an investigation in the course of which they examined 2075 autopsy specimens and studied six normal females as 'experimental animals'.[270] They found corpora lutea only in the ovaries of pregnant animals and found no evidence of spontaneous ovulation by rectal palpation of the ovaries of the 'experimental' animals. They concluded that ovulation only occurs after copulation, but they confused the issue somewhat by stating that they found the species to be 'polyoestrous'. By this term, however, they apparently referred to their observation that 'If the oestrous camel is not mated, the mature follicle undergoes atresia and ovulation does not take place'. The oestrous condition is presumably interrupted during the period of decline of one follicle and the maturation of its successor. These periods of anoestrus appear, from this description, to be more clearly defined than those in the ferret or the cat. In these species also it is probable that the ripe follicles are replaced at intervals, but they are polytocous and the corpora lutea of each set do not necessarily regress simultaneously, so that oestrus may be more or less continuous.

The New World relatives of the camel and dromedary, the llama and alpaca, also appear to be 'induced ovulators'.[120, 135] The llama (*Lama glama*) is a transport animal, the alpaca (*Lama pacos*) is grown for its wool; both are thought to be descended from the wild guanaco (*Vicugna guanicoe*). With the smaller wild vicuna (*Vicugna vicugna*), whose wool is even more valuable, these species are of considerable economic importance in the cordilleran countries of Argentina, Chile and, especially, Peru and Bolivia.

Llamas breed from December to May (midsummer to late autumn) but less intensively from February onwards. Their low conception rate, and consequently low fertility, presents a problem of practical importance in Peru and Bolivia. When conception fails after mating and ovulation the corpus luteum regresses in about two weeks and the animal enters oestrus again. Ovulation has been found to be fairly easily induced by the administration of LH, and it sometimes occurs spontaneously (i.e., without mating) in the height of the breeding season.

Spontaneous ovulation

Spontaneous ovulation with 'inactive' corpora lutea and consequent absence of a luteal phase in the 'unmated' cycle, appears to be characteristic of murine rodents. Such short oestrous cycles have so far been described only in the mouse,[3, 281] the rat[228, 240] the Chinese hamster[282] and the golden hamster.[92] Just as spontaneous ovulation may sometimes occur in a species that normally ovulates only after coitus, the converse may be observed in that coitus may induce or hasten ovulation in a species where it normally occurs spontaneously. This has been described in the rat[17, 18, 19] and it is thought to occur in man.[114] It has also been noticed in zebras in captivity.[183]

The golden or Syrian hamster (*Mesocricetus auratus*) is now widely used in biological and medical research, although the laboratory population does not approach the numerical strength of the rats and mice. All the stocks apparently derive from three litter-mates, one male and two female, captured near Aleppo in 1930. The first description of their reproductive physiology was published in 1934.[52] In 1968 a comprehensive monograph was devoted to the use of the golden hamster as a laboratory animal and with regard to its reproduction says 'It is erroneous to assume that the golden hamster resembles the rat, mouse or any other rodent with respect to any facet of physiology'.[212] This may be regarded as a somewhat partisan view; the hamster is decidedly rat- and mouse-like in that it has a 4-day polyoestrous cycle, pseudopregnancy is induced by coitus, deciduomata can be induced by trauma and vaginal smears can be used to follow the cycle. Prolactin is an essential component of the luteotrophic complex, there is an immediately postpartum heat, ovarian progesterone is required throughout gestation, and hysterectomy prolongs the duration of pseudopregnancy from about 9 or 10 days to about 16 days, the duration of pregnancy.

It may be that 'short' cycles, in unmated animals, would be more frequently encountered if more species were maintained in captivity. Perhaps they are noticed in murine rodents because several species of them are kept in the laboratory or as pets. More attention is now being given to breeding the less 'common' animals in captivity than was formerly the case. This is partly because it offers a means of further discovery and partly because the demand for exhibits in zoos and menageries has become so great that the supply from the wild is in danger. One common British mammal that is easily captured and kept in captivity is the hedgehog, and there seems little doubt that it ovulates spontaneously or that the corpora lutea are relatively inactive in the absence of mating. Curiously, there appears to have been no detailed investigation of the reproductive cycle of this species since that by Deanesly in 1934,[91] possibly because her account seemed to leave little to be done. The embryology of the hedgehog and particularly the transmission of passive immunity from mother to foetus, has been described in a series of papers by Morris, who also made incidental but valuable observations

on the breeding season and on the rearing of young hedgehogs.[262]
Deanesly's material was derived from 'hedgehogs taken straight from
the field, since the condition of the reproductive organs may become
affected after only a few days in captivity'. However, some that were
taken in February and March, while still anoestrous, 'In due course,
showed breeding season development of the reproductive organs and
ovulated spontaneously and normally in captivity'. It would seem,
therefore, that there is scope for the exploration of this species as a
laboratory animal.

The mole is another common British insectivore the ovarian cycle
of which has been studied mainly from material collected in the wild.[96]
The female bears a single litter of three or four young in spring, enters
lactation without a post-partum oestrus and then remains anoestrous
until the following spring. The ovaries of the mole are characterized by
the large amount of 'interstitial' tissue that develops during the anoestrous period. This tissue is of medullary origin and therefore is homologous with testis tissue, which it resembles, although no germinal
differentiation occurs in it. During anoestrus the follicular tissue is
confined to a small region near the surface of the ovary, and the interstitial tissue, consisting of tubules in a matrix of eosinophil cells, becomes
predominant. With the approach of the breeding season the follicles
enlarge, and during pregnancy well defined corpora lutea are formed.
The testis-like appearance of the ovary during anoestrus appears to be
responsible for the belief that the mole changes sex during the year, or
that all those found in winter are males. It is possible, but not known,
that ovulation depends on copulation in the mole. This uncertainty is
yet another illustration of the limitations imposed by the difficulty of
breeding such animals in captivity, but it is possible that the difficulty
is more apparent than real, since few people who are in a position to
utilize such a colony also have the facilities and leisure to establish
one.

An 'active' luteal phase follows spontaneous ovulation in the guinea-
pig (alone among the common laboratory animals), in the dog and in all
the common farm animals (cow, sheep, pig, horse and goat). The farm
animals, and the guinea-pig, are polyoestrous—that is, they experience
a succession of oestrous cycles during the breeding season, or throughout the year, unless pregnancy supervenes. The dog is unusual in that
the female is monoestrous; if the bitch is not mated she will enter a
period of sexual inactivity (anoestrus) after a luteal phase similar in

duration to that of pregnancy, about two months. The interval between successive 'seasons' varies with the breed and with the individual animal. It was recorded as 7 to 8 months in a colony of foxhounds, airedales and mongrels; oestrus usually lasted 10 to 12 days, but it ranged from 6 to 24 days.[164] The prolonged period of heat and the monoestrous habit are reminiscent of the 'induced ovulators' among the Carnivora. In some respects the dog is intermediate between this condition and the typical polyoestrous species. The dog is also unusual in that a periodic discharge of blood is associated with the cycle. The blood is of uterine origin, but it is distinct in character from the menstrual flow of the higher primates. It is discharged before oestrus, it does not involve the shedding of uterine lining, and it is apparently caused by a high concentration of circulating oestrogen; it is certainly not caused by, and cannot be experimentally induced by, progesterone withdrawal.

Among the farm animals, the duration of the unmated cycle is 16 days in the sheep, and 21 days in the cow, pig, horse and goat. The luteal phase of the cycle is relatively shorter, and the follicular phase longer in the horse than in the other species. This may conceivably be related to the superficial type of placentation in the horse and to the very large size of the mature follicle, which reaches a maximum diameter of about 7 cm and holds more than 100 ml of fluid.

The guinea-pig ovulates at 16-day intervals. During the 10 days before ovulation, follicles mature in rapid succession. Most of them undergo atresia, but the development of others coincides with the approach of oestrus, and these mature in an endocrine environment that allows, or causes, their ovulation. The oestrous cycle of the guinea-pig was one of the first to be followed in detail, and mention has already been made of the vaginal cytology, as described by Stockard and Papanicolaou in 1917.[368] Later, in considering the regulation of the periodicity of the cycle, and of ovarian function during pregnancy, we shall have occasion to refer to the pioneer work on the guinea-pig by Loeb, who described the cyclic changes in the ovary as early as 1911. In scanning the literature of the succeeding decades one gets the impression that the guinea-pig was soon superseded by the rat as the favourite laboratory animal in this field. This was undoubtedly due to the convenience of the short cycle of the rat in many endocrinological studies, and to its superior fecundity. The average litter size of the rat is about 10, and its gestation period is 21 days; the corresponding figures for the guinea-pig are approximately 3 and 65. In more recent years the guinea-pig has perhaps been increasingly used, especially in work bearing on the control of the luteal phase.

Detailed information about the oestrous cycle of animals in the wild state is still confined to relatively few species although, of course, something is known about a great number of them. The ovarian cycle of the wild brown rat[290] appears to be similar to that of the laboratory rat, but it may breed more seasonally in some habitats than in others; in the laboratory it breeds throughout the year. Information about the wild relatives of the guinea-pig, the cavies of South America, is extremely scanty. This situation is being remedied, however, by the establishment of breeding colonies of a number of hystricomorph rodents, collected in South America, in the Wellcome Institute of Comparative Physiology in London. Projects of this kind have been rendered feasible, in recent years, by the great development of rapid communications, especially by air. The Hystricomorpha are very diverse, even bizarre, in reproductive pattern. In most of them the gestation period is long in relation to body size; in the mountain viscacha only the right ovary, apparently, ovulates and forms corpora lutea, and pregnancy is almost invariably in the right uterine horn. Accessory corpora lutea are formed and accumulate throughout pregnancy, by follicular atresia in the same ovary.[289] The plains viscacha, on the other hand, acquires a huge number of small accessory corpora lutea by extraordinary multiple ovulation.[387] Both these species are referred to in a later chapter (p. 129).

A wild animal rarely fails to mate at oestrus, so that the 'basic' cycle may not be encountered. In addition, the estimation of time intervals such as the length of the cycle, or even the gestation period, is difficult. For example, in a study of the African elephant I examined the carcasses of 68 adult females. Thirty-one were pregnant, and only three had what appeared to be relatively recent corpora lutea. On the basis of this material I suggested that fertile mating, at puberty or after anoestrus, is preceded by one or two infertile ovulatory cycles in this species.[291] It is impossible to determine, from material of this kind, whether or not such ovulations are accompanied by oestrus and mating. More recently, Short has arrived at a similar interpretation in the light of his own observations in Africa,[353] but whereas I concluded that *Loxodonta* is polyovular, Short's material suggested that it is monovular. Similarly, in spite of the impressive amount (and quality) of work done in recent years on the ecology of the African elephant, the only reliable information about its gestation period is derived from a few specimens bred in captivity. One was born in Basle zoo after a gestation period of between 649 and 661 days, the mother having been served by two bulls 'at intervals' over a period of 12 days.[215] It is unlikely that the cow elephant would remain in oestrus for 12 days, in the wild state, but this remains the best available evidence about the gestation period.

An intriguing problem has arisen in the study of elephant reproduction, Dr. Short having found it impossible to demonstrate the presence of progesterone in the peripheral blood taken from an African elephant in early pregnancy, with a normal foetus of about 30 mm crown-rump length and a corpus luteum of normal appearance. Apart from the unlikely possibility that the steroid was eliminated from the blood in the brief post-mortem interval before the samples were taken, it is difficult to account for this finding except by assuming that the corpus luteum of the elephant, alone among the mammalian species so far studied, secretes a progestagen other than progesterone.

The establishment of the great National Parks in East Africa has made it possible to study, as well as to conserve, the large mammals that are among their main attractions. The laboratory of the Nuffield Unit of Tropical Animal Ecology, in the Queen Elizabeth National Park, Uganda, is well placed for work of this kind, and the most abundant material, as well as the most urgent problem, was provided by the hippopotamus population in the immediate vicinity of the laboratory. Within a few years of the establishment of the National Parks the density of the hippo population in this area led to alarming deterioration in the environment, and a policy of 'cropping' was adopted. During two years work under this policy, the zoologists examined 2,000 hippo carcasses.[218] The carcasses were processed for food on the spot, so the investigators had to work quickly.

The general structure of the female hippo reproductive organs somewhat resembles that of the pig, with a long cervix with interlocking transverse ridges. The placenta, also like that of the pig, is epithelio-chorial and diffuse. The hippo, however, is monotocous, and only a single follicle develops except in rare cases of twins. A single large corpus luteum is formed at ovulation but, as pregnancy progresses, accessory corpora lutea are formed in both ovaries, some by the luteinisation of atretic follicles and others, probably, from follicles that ovulate.

The ovarian cycle of the other great pachyderm of Africa, the rhinoceros, has not yet been studied in detail, nor has that of the two-horned Asian species. The African antelopes have received considerable attention, mainly centred on their breeding season and ecology (see p. 160 *et seq.*).

Among the Artiodactyla the reproductive physiology of the deer (Cervidae) is interesting by reason of the relation between the gonadal hormones and antler growth and dehiscence. Seven species of deer now breed in Britain, either as native wild species or as introduced species living in wild conditions or as domesticated or 'park' deer. The red, roe, fallow and reindeer are or were indigenous species, but it is probable that the fallow deer became extinct here with the last glaciation and was subsequently re-introduced. The reindeer is thought to have survived in Britain until about 1300 A.D. and it was reintroduced in 1952. The other introduced species are the sika (*Cervus nippon*), the Chinese water deer (*Hydropotes inermis*) and the muntjac or barking deer (*Muntiacus sp.*). The Chinese water deer is exceptional in that neither sex grows antlers; in the reindeer both sexes do so, and in the remaining species mentioned only the males have antlers. The growth and the casting of the antlers in the male, of the red deer at least (p. 185), is controlled by the testis steroid hormone, testosterone. A seasonal (or pubertal) increase in testosterone secretion causes the shedding of the 'velvet', and a seasonal decline in this hormone causes the antlers to be cast.[224] The female reindeer casts its antlers after calving in the spring, having retained them for some months longer than the male. It is not clear what controls antler growth in the female reindeer, but it is evident that the antlers are retained while the corpus luteum of pregnancy is functional and are cast when it regresses, which suggests that progesterone is involved.

Roe deer are commonly regarded as monoestrous animals—that is, experiencing only one oestrous period during each mating season, effectively one a year. This may be a false impression, due to the high fertility of the species, such that the first oestrus of the season almost invariably leads to pregnancy, which lasts for 40 weeks.[67] This long gestation period is associated with 'delayed' implantation (see p. 118).

'*Silent heat*'. The occurrence of infertile cycles at puberty, or at the onset of the breeding season as suggested (above) in the case of *Loxodonta*, may be a widespread phenomenon. It has been described, for example, in the Uganda kob[55] and it has long been recognised in the sheep,[147] where one or more ovulations usually occur without oestrus— hence the term 'silent heat'—at the beginning of the breeding season. It is said to occur frequently in the goat,[87] rarely if at all in the cow[160] or sow.[57] The mare may ovulate regularly without showing signs of heat.[237]

It has been suggested that a small amount of progesterone, as well as oestrogen, is required to induce oestrus, and that a 'silent heat' may initiate a breeding season because ovulation then occurs in an ovary which is secreting no progesterone. The waning corpus luteum of the 'silent heat' is envisaged as providing a sufficient amount of progesterone[323] or as conditioning the animal so that oestrus accompanies a subsequent ovulation.[313] 'Silent heat' may also occur near the close of the breeding season but in this case there is probably a deficiency of oestrogen.

Recent work has shown that there is a pre-ovulatory rise in the concentration of circulating (plasma) progesterone in a wide variety of species. This is true of rat, guinea-pig, man, and the domestic sheep, pig and cow, in all of which it evidently facilitates ovulation. This notwithstanding the fact that progesterone, administered systemically or orally, prevents ovulation and suppresses cyclic ovarian activity, as in human contraceptive practice and in the use of progesterone or synthetic progestagens to defer ovulation in other species in order to synchronize oestrus when the steroid is withdrawn.[302] The contrasting effects depend on the phase of the oestrous cycle at the time progesterone is administered. By the term 'facilitation' is implied that the action of progesterone in this instance is not to cause ovulation directly but to increase the animal's sensitivity to some other ovulatory stimulus. Thus, in earlier experiments by Döcke and Dörner, it was found that progesterone, administered in the right dose at the right time, increased the proportion of rats that ovulated in response to oestradiol benzoate. Ovulation was inhibited if the progesterone was administered within 3 days of the oestrogen, but it was facilitated by doses within a range of 0·5 to 2·5 mg, given 3 days after the oestrogen. Progesterone itself did not induce ovulation. In their later paper these authors added to existing evidence that this action of progesterone is exerted through the hypothalamus, and found that hypothalamic implants of progesterone had greatest facilitatory effect when they were implanted in the ventromedial arcuate region. They suggested that its effect was to lower the activation threshold of receptors in this region to stimuli thought to reach it from the medial pre-optic area of the hypothalamus.[103]

Hibernating bats. A type of reproductive cycle in which follicular development is arrested, and spermatozoa are stored, throughout a long hibernation, is entirely peculiar to some species of the Microchiroptera. They copulate in the autumn, when the ovaries contain a single vesicular follicle which has an unusually high concentration of glycogen in the cumulus cells.[390] The maturation of this follicle is arrested during hibernation, but when normal metabolic activity is resumed in the spring its growth re-commences, and is so accelerated that ovulation occurs within 24 hours. The oocyte from this 'delayed action' follicle is fertilized either by a spermatozoon from a second copulation (in the spring) or, in some cases, by one of the spermatozoa that have been held in the female genital tract throughout the hibernation period.[252] Presumably, if mating were somehow prevented, ovulation would recur, but this does not appear to have been observed. This extraordinarily protracted follicular phase, peculiarly associated with hibernation, may be regarded as an adaptive modification of the reproductive cycle, comparable in kind with the delay of implantation by means of which an 'embryonic diapause' is interposed in the development of some species. The effect is to adapt the temporal relations of the reproductive or developmental process to the exigencies of the animal's way of life.

4: The Internal Control of Ovarian Periodicity

The special position of the corpus luteum

The follicular phase of the cycle is relatively constant in duration in different species, being governed only by the time taken for a follicle, or a group of follicles, to grow to mature size. In animals with an active luteal phase, further ovulation is delayed until the corpus luteum regresses. The repetition of the cycle, and its periodicity, therefore depend upon whatever mechanism causes this regression, and this mechanism may be regarded as a kind of pacemaker in the cyclic sequence of events. Furthermore, the luteal phase is prolonged during pregnancy and the corpus luteum plays a vital part in retaining the embryo in the uterus. The mechanism controlling its persistence is therefore of paramount importance throughout the animal's reproductive life. There are many eutherian species in which the corpus luteum does not last throughout pregnancy (see p. 117), but none in which its duration is shorter in pregnancy than in the unmated animal. At present we are considering the sequence of events in the non-pregnant animal.

The corpus luteum, as a dominant factor in the reproductive cycle, is peculiar to the Mammalia. In elasmobranch fishes and in the Amphibia, the epithelium of the discharged follicle becomes 'luteinized' but there is no evidence that the bodies so formed have any function. Again, among the reptiles, a well defined corpus luteum is formed in many species, but its occurrence is not clearly related to the reproductive habit of the species. Viviparity involving retention of the embryo in the uterus (oviduct) is found among members of all these groups, but its distribution does not coincide with that of post-ovulatory luteinization. True viviparity is found in some species of teleost fish, but placentation is always intra-ovarian, and the post-ovulatory follicle probably has no special function although 'pre-ovulatory corpora lutea', which are derived from atretic follicles, appear to have some function.[45] The alternative term 'corpus atreticum' is more appropriate.[65, 196]

We have already attributed the function of inducing ovulation to the 'luteinizing' hormone, LH, and the formation of a corpus luteum is apparently an inevitable consequence of ovulation throughout the mammals. The existence of a third gonadotrophin, controlling the survival and function of the corpus luteum after its formation, was apparently first postulated by Astwood and Fevold in 1939.[22] They showed that progesterone inhibited corpus luteum formation in the rat but did not cause the regression of established corpora lutea. They also pointed out that LH was inhibited at precisely the time when luteal function could be expected to be maximal, and quoted Greep as having shown that LH could cause luteal regression. They concluded that 'The pituitary factor responsible for the maintenance of luteal function is probably distinct from luteinizing hormone but as yet the identity of this corpus luteum activator has not been established'. The terms 'luteotrophic hormone', and 'luteotrophin' were introduced by Astwood in 1941.[21] Luteotrophin (or in present American usage, luteotropin; LTH) was therefore postulated as the pituitary factor responsible for the continued existence, and function, of the corpus luteum. It has since become clear that we may have to consider as distinct phenomena the survival of the corpus luteum as a histological entity, its function in terms of the synthesis of progesterone, and its function in terms of the secretion of progesterone, but this is a complication to which we shall return later. The identity of LTH has received much attention, and it seems clear that its nature, and its importance, varies from one species to another. To summarize for the sake of clarity: it is probable that prolactin acts as the luteotrophin in the rat and the mouse, and that LH has this function, as well as being the ovulating hormone, in some species, including the cow. Prolactin, or lactogenic hormone, which elicits lactation in mammals, is universally present in mammalian pituitary glands, as is LH.

Cyclic events in the ovary could be explained without invoking any further hormonal factors, supposing that luteal regression would follow the withdrawal or suppression of LTH, and supposing the existence of suitable feedback from the ovary, controlling the FSH/LH balance at appropriate times. Pregnancy, in such a scheme, would prolong luteal maintenance by stimulating LTH as a result of uterine (decidual) changes consequent upon the implantation of the embryo.

It has in fact been shown that feedback mechanisms of the required sort do exist, as indicated in Fig. 2. They operate not directly on the pituitary gland but upon centres in the hypothalamus that produce the gonadotrophin-releasing factors which control the pituitary secretion. However, attention has recently been redirected to the role of the uterus in luteal regression in the non-pregnant animal. The evidence is examined in some detail in a later section (p. 102).

Maintenance, function and regression of the corpus luteum

We must now proceed to examine in greater detail the evidence concerning the function of the corpus luteum, and its maintenance and regression. It is a subject involving a very wide area of the field of reproductive physiology, and a complete synthesis of the mass of available experimental data is beyond the scope of this volume. Indeed, a complete synthesis is impossible in the present state of knowledge, because many of the observations appear to be contradictory. The discrepancies will be explained when the mechanisms are thoroughly understood, but like many areas of biology, endocrinology is still in the stage when a new observation is at least as likely to confuse as to illuminate. It has been despairingly said that any hormone can have any effect, according to the species, the dose level, the condition of the animal and the 'endocrine environment' in which the hormone is applied. An attempt will be made, however, to present a clear picture of what is known of the physiological mechanism as a whole, and to evaluate some of the relevant information about its components.

The very generalized diagram in Fig. 2 is intended merely to provide reference points in the description that follows. It would be impossible to include in it all the effects that the various hormones have been shown to exert, and no single schema would be appropriate for all species, but the basic events are clear.

Model control systems

Descriptions of observed effects have often been presented in diagram form, in varying degrees of complexity according to the range of information included. Such diagrams are sometimes referred to as 'models', but they are by their nature static. A diagram of this kind is apt to be more valuable to its compiler than to anyone else, and the compilation itself presents a useful heuristic exercise. Perhaps the best diagrams of this type are those so constructed that they could serve as plans, or circuit-diagrams, for a dynamic model capable of demonstrating the effects of raising or lowering the various inputs, and the possible interactions of various stimuli in the multiple control of the cycle. All such 'models' are intended to advance 'the conceptualization of functional relationships'.[340]

LH as ovulating hormone

In tracing the various pathways involved, the induction of ovulation by pituitary LH provides a good starting-point, since it appears to be universal throughout the mammals as well as being a key point in the ovarian cycle whatever other manifestations accompany it. Parlow's ovarian ascorbic acid depletion test[286], although it is now being superseded by radio-immuno-assays has been a valuable specific and quantitative means of estimating LH, and its use has demonstrated a pre-ovulatory rise in this hormone in the plasma of cyclic rats[339] and a drop in pituitary LH content at the time of ovulation.[198] LH synergizes the effect of FSH in stimulating follicular growth and oestrogen secretion, but ovulation appears to depend on a sudden and transient increase in the amount of LH released by the pituitary. The classic work of Fee and Parkes[130] showed that the release of hormones from the pituitary gland to provide the ovulatory stimulus lasted about 60 minutes in the rabbit, and Everett[125] showed that it lasts about half as long in the rat. Many mammals, including those which ovulate spontaneously, appear to resemble the rabbit in that ovulation occurs approximately 10 hours after this LH release. The luteinizing hormone itself is metabolized and removed from the circulation in much less time than this, so that its effect must be to initiate some process which thenceforth runs its course to terminate in ovulation.[268]

The mechanics of follicular rupture, as far as they are at present understood, have been outlined earlier (p. 25). It is clear that the actual process of rupture involves a very localized area of the follicular wall at the surface of the ovary, but the process of luteinization involves the whole circumference of the follicle, and while the thecal capillaries contract over the rupture area, and appear to withdraw from it, they proliferate throughout the remainder of the theca and soon penetrate the membrana propria. With the expulsion of the liquor folliculi, bearing the egg and its surrounding cumulus, the follicle temporarily collapses, but its cavity is soon refilled with 'lutein' tissue, and blood capillaries ramify throughout the solid corpus luteum. All these changes, so markedly different over the apex of the follicle, and involving the cytological and endocrinological transformation of both theca and granulosa, apparently follow inevitably in the train of the brief ovulating stimulus.

It may be useful, at this point, to summarize the hormonal changes leading up to ovulation in animals such as the cow and sheep, in which ovulation does not depend on coitus, and prolactin does not play so dominant a part in maintaining luteal function as it does in the rat and mouse. The concentration of gonadotrophins in the pituitary gland of the cow has been measured at intervals throughout the oestrous cycle. The FSH content rises fairly steadily until about day 18, then falls slightly, then rises steeply for about two days, approximately during days 18 and 19, and falls precipitately just before ovulation.[152] In both cases the sudden fall before ovulation coincides with a 'surge' or transient rise in the circulating plasma level of the hormone.

Figure 11 is an attempt to piece together the story as it might apply to the sheep (in purely pictorial terms, no significance being attached to the ordinal values!) in the days immediately before and after oestrus, taking the information from various sources. The release of FSH from the pituitary, and the consequent rise in circulating levels, brings about the final growth of the ripening follicles and their peak production of oestrogen. This brings the animal into heat. During oestrus the pituitary, in which LH has been rapidly accumulating for some days, suddenly releases this hormone, providing the ovulatory peak, so that ovulation occurs at or near the end of oestrus. During the final maturation of the follicles they produce a little progesterone, which augments the dwindling supply from the regressing corpus luteum and appears to 'facilitate' ovulation. Prolactin, which has been steadily accumulating in the pituitary, is released after ovulation and probably plays some part in maintaining luteal function (i.e., is luteotrophic) in all species. As will appear later, its luteotrophic activity is more evident in the sheep than in the cow or pig.

It will be seen from Fig. 11 that the typical oestrous cycle does not consist of alternating 'phases' of progesterone and oestrogen domination. Significant circulating levels of oestrogen are found, in the sheep, only over a very brief period, and they in fact reach a maximum at the time of oestrus and then immediately start steeply to decline. If the primate (menstrual) cycle is to be derived from the pattern shown here, the principal change would be a delay of follicular maturation for at least a week after the decline in plasma progesterone levels. In effect, since progesterone is rapidly metabolized, this decline is more or less co-incident with the regression of the corpus luteum. Such a separation of progesterone and oestrogen peaks would result in a relatively prolonged cycle of a type that has already been described (Fig. 10). On the face

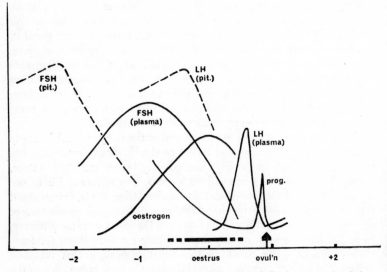

Fig. 11. Diagrammatic representation of some of the hormone 'peaks' near the time of oestrus in the sheep.

of it, one might be tempted to speculate that the overt manifestation of oestrus is possibly prevented at the time of ovulation, in the middle of the menstrual cycle, because the oestrogen is not synergized with pro-gesterone—and that menstruation is perhaps prevented at the time of ovulation in the oestrous cycle because oestrogen is present when the progesterone is withdrawn. The evidence from experimental admini-stration of the two steroids does not support this facile explanation, however, and we have already seen that recent work has shown that ovulation and menstruation can almost coincide in the tupaiids (p. 76).

Even if it were possible to 'derive' one type of cycle from another, the exercise is likely to be useful only in so far as it provides a logical framework in which to compare, and perhaps to explain, the kinds of mechanism involved. To attribute phylogenetic significance to the process would be quite unwarrantable.

Luteotrophin

Once the corpus luteum is established and functional, the question arises whether a further gonadotrophin is required for its maintenance. Such a gonadotrophin would be by definition a 'luteotrophin'. Later in the cycle, when it becomes essential to stop the secretion of progesterone, a further question arises: is this brought about by cutting off the luteotrophin, or does it require a further stimulus which, if it proved to be humoral in nature, could be termed a 'luteolysin'? In examining these questions, we must bear in mind that it is unwise to assume that a corpus luteum is synthesizing steroids simply because it is seen to be present in the ovary, whatever its histological appearance.

Rat and mouse: role of prolactin as luteotrophin

Both the persistence of the corpus luteum and its ability to secrete progesterone depend on a luteotrophic hormone in the pseudopregnant or pregnant rat and mouse. This luteotrophin was identified with prolactin (lactogenic hormone) in the mouse[110] and in the rat.[21, 123] A considerable amount of evidence has since accumulated which seems to show that this condition does not apply to mammalian species in general, although it is possible that prolactin contributes to a complex control mechanism or 'luteotrophic process' in which no single substance can be specified as 'the' luteotrophin. Rothchild, in discussing 'The nature of the luteotrophic process',[316] maintains that prolactin may well prove to play a part in the maintenance of luteal function in all mammals. He appears to deprecate the kind of reasoning that leads to the conclusion that prolactin is not luteotrophic because, for example, its administration fails to prolong the oestrous interval, or to prevent luteal regression in circumstances in which it normally occurs. He maintains that the possibility that prolactin plays some part in luteal maintenance has not been conclusively excluded in any instance, and warns that 'the exact nature of the luteotrophic process will probably not be known until the problem is approached with the sophistication that its complexity requires'. One might quarrel with the last statement on purely logical grounds since, when the 'exact nature' of a phenomenon is known, the methods of its elucidation have necessarily proved themselves adequate. The warning is clear, however, and it is indisputable that the administration of additional exogenous hormone to an intact organism is of limited value, or that the role of a hormone in nature can only be understood when its synergisms, and the conditions in which it functions, are taken into account.

The question of synergism is particularly important in the case of prolactin. In particular, 'The problem of the relationship between prolactin and growth hormone in animals is highly complex'.[15] The two hormones have been extracted from sheep and cow pituitaries as apparently separate chemical entities, yet a rabbit anti-serum to human growth hormone (HGH) was found to inhibit the 'prolactin' activity of human pituitary extracts. This led to the suggestion that, in man, prolactin and growth hormone are identical. More recently, however, tissue culture experiments have provided results that seem to imply the existence of two cell lines in human foetal pituitary tissue, one elaborating GH and the other, prolactin. The separate identity of human prolactin in the normal state has yet to be unequivocally proved, but once again the distinction between its effects and those of growth hormone seems likely to be one of degree. Growth hormone exerts some prolactin-like activity, and prolactin exerts some GH-like activity, and the degree of similarity, and of synergism, probably varies widely from species to species. It may be the case that, in any species, the more prolactin resembles GH, the less it will resemble LH. It is not probable that the complexities of function of these hormones will be explained in the immediate future, and in the meantime biological problems such as those of corpus luteum maintenance have to be handled conceptually, using terms such as 'the luteotrophic process', which we have already introduced above.

Prolactin was identified as the luteotrophic hormone in the rat because the corpora lutea of the unmated rat behave like those of the pseudopregnant rat (and like those of the non-pregnant guinea-pig) when prolactin is administered to the intact animal. Although the term 'luteotrophin' has been criticized—Rothchild says that 'Riddle[312] has perhaps rightly (if somewhat acidly) criticized it'—it is clear that prolactin plays a very prominent part in luteal maintenance in the rat and mouse, and a much less prominent, perhaps negligible, one in many other species. Rothchild poses the problem: 'If the luteotrophic hormone is different in different mammals, what accounts for the striking similarity of luteal morphology and function in most mammals? On the other hand, if the luteotrophic hormone is really the same in most or all mammals, what accounts for the failure to demonstrate a luteotrophic effect of prolactin except in the rat and mouse?' However, few, if any, phenomena in the normal physiology of reproduction depend solely on the action of a single hormone, and it is probable that the completely normal response is to a complex of hormones, in which the proportions of the constituents vary from time to time and from species to species. The argument that the similarity of the 'luteal' tissues in different species points to a common stimulating hormone, seems to me to carry little weight. The nature of a cell's response to a stimulus, if it occurs at all, is likely to be governed by the nature of the cell rather than the nature of the stimulus. An interesting example of this has recently been demonstrated experimentally.[384] The ovaries of rats hypophysectomized very early in pregnancy were stimulated to secrete sufficient oestrogen to allow implantation (the oestrogen 'surge'—see p. 123) by the subcutaneous injection of various protein extracts. The cellular activity involved was presumably similar in kind to that which is evoked by pituitary gonadotrophins in the normal course of events.

Having posed the problem, Rothchild proceeds to examine it, first summarizing the evidence that prolactin is not luteotrophic. The evidence is derived from a wide range of researches, all relating to the control of the ovarian cycle, and the references are cited in Rothchild's article. In brief:

1. The administration of prolactin does not prolong the activity or existence of the corpora lutea in the intact animal, in species where the corpora lutea are normally active after spontaneous ovulation. This applies to the human species among many others. Menstruation can be delayed by treatment with progesterone or with chorionic gonadotrophin (containing LH), but not by prolactin. However, abnormal persistence of the corpus luteum, causing amenorrhea, is often accompanied by galactorrhea, suggesting that an excess of prolactin is involved.

2. Luteal regression following hypophysectomy is not prevented by prolactin, but may be delayed by other hormones. The persistence of the corpus luteum of the sheep after pituitary stalk section, as distinct from hypophysectomy, is referred to below (p. 99) and it will be seen

that there is convincing evidence that prolactin is an important luteotrophic hormone in this species.

3. Prolactin does not increase the rate of production of progesterone by luteal tissue *in vivo* or *in vitro*, whereas other gonadotrophins have this effect.

4. In some species, the ovaries contain small and relatively inactive corpora lutea during lactation, although prolactin is necessarily being secreted.

Rothchild points out sundry weaknesses in what one might call the case against prolactin as luteotrophin. It has been argued, for instance, that the failure of prolactin to induce progesterone secretion by the ovary of the oestrous rabbit supports this 'case'. In fact, of course, 'Even in the rat, prolactin does not stimulate the secretion of progesterone by an ovary that lacks corpora lutea'. There is perhaps an element of contention in Rothchild's argument, since he seems to seek a feasible possible role for prolactin in the luteotrophic process, whether or not the available evidence demands it. The argument is justified, however, by the problems that arise if one is forced to conclude that this process is fundamentally different in various species. To avoid this conclusion, or to set it aside until the process is more fully understood, Rothchild suggests that prolactin may have a subsidiary role in the luteotrophic process in many mammals, and a dominant but not necessarily exclusive role in the rat and mouse. This is in line with what was said above, of a possible 'complex' of factors, and it is the basis for writing of a 'luteotrophic process' rather than a 'luteotrophin'. We meet the same conflict between caution and simplification in connection with the regression of the corpus luteum ('luteolysin' or a 'luteolytic mechanism'—see p. 102). Whereas the corpus luteum of the unmated rat persists for only a few days, and that of the pseudopregnant rat does not synthesize progesterone in significant amounts in the absence of the pituitary gland and therefore of prolactin, the corpus luteum of some species has been found to grow and apparently function after hypophysectomy early in the cycle. This has been noted in the guinea-pig in the course of experiments on follicular growth.[99, 315] The effects of hypophysectomy at different stages of the cycle, in pregnant and in hysterectomized guinea-pigs, have since been studied in more detail,[179, 320] and we shall refer to these observations again later in discussing luteal regression. Persistence of the corpus luteum after hypophysectomy has also been observed in pigs[113] and in sheep.[101] Clinical endocrinologists suspect that the human corpus luteum may persist independently of pituitary stimula-tion[382] but no direct evidence of this is available in man or in any species other than those already mentioned. Of course, the survival of the corpus luteum after hypophysectomy does not necessarily mean that pituitary hormones play no part in regulating its activity in the normal intact animal. Similarly, prolongation of luteal function by the administration of a hormone does not necessarily indicate that the hormone normally does this. Further work on the sheep showed that the corpus luteum secreted normal amounts of progesterone for only about 9 days after hypophysectomy, but did so for the normal duration of the cycle after pituitary stalk section.[100] From other investigations it is probable that the pituitary gland will continue to secrete prolactin after severance of its connection with the hypothalamus, but will no longer secrete FSH or LH, so the authors drew the inference that prolactin is necessary to maintain luteal function (progesterone secretion) for the normal period.

In the case of the guinea-pig it was found that corpora lutea persisted for at least 5 weeks after hypophysectomy performed 2–3 days after ovulation, but the average number found in the experimental animals at autopsy was about half the average number in control animals. The corpora lutea that survived reached approximately normal size, but took longer to do so. Stages of regression were not observed, and one is led to conclude that about half the corpora lutea regressed rapidly within a short time after hypophysectomy.[179] Why some should survive and grow, while others in the same ovaries regress and disappear, remains a mystery. Those that survived contained approximately normal amounts of progesterone and subsequent work has shown that the luteal ultra-structure was normal. As in the case of the sheep, therefore, corpora lutea can persist in the guinea-pig in the absence of the pituitary gland, but the normal course of events is disturbed. In the sheep this appears to be because prolactin is missing; the pituitary 'regulator' in the guinea-pig has not yet been identified*.

The experimental approach to a particular problem in different animals is governed to some extent by the characteristics and particu-larly by the size, of the animal itself. It is only in relatively recent years that hypophysectomy has been a practicable experimental method for use on sheep and pigs, and it has not, at the time of writing, been applied to work on the cow, although a method of pituitary stalk section has been developed. Convincing evidence about luteal maintenance in cattle has

*See p. 35 (footnote).

nevertheless been obtained; it very strongly suggests that LH can be regarded as the bovine LTH. An outstanding contribution to this subject has been made over the past decade by Hansel and his collaborators at Cornell University.[165]

Far from being daunted by the size of the cow for laboratory purposes these authors claim that it 'has proved an excellent experimental animal for studies on maintenance of the corpus luteum'. One advantage is that, because of the animal's size, its ovary can be palpated *per rectum* and the corpus luteum can be located, and even extirpated, in this way —a technique exploited many years ago.[159] The American workers have shown that an ovarian biopsy can be performed very conveniently through a small incision in the anterior vaginal wall, and they have used this as a means of obtaining luteal tissue of known age for the study of the effects of various gonadotrophins on progesterone synthesis *in vitro*. They have also shown that normal luteal development can be inhibited by the administration of oxytocin during a period from the second to the sixth day of the cycle, and this has proved a most useful technique. The mechanism of the effect, which is apparently peculiar to the cow, is obscure; oxytocin causes the release of a gonadotrophin from the pituitary of both sexes of several species, and it is possible that in the cow it exhausts the supply to an extent that irreversibly inhibits luteal growth and function when it is done at a critical time.

Preparations containing LH overcame the inhibitory effect of oxytocin administration, whereas other pituitary hormones (prolactin, FSH and growth hormone) did not. Purified bovine LH, administered in mid-cycle, significantly prolonged the functional life-span of the corpus luteum; prolactin had earlier been shown not to have this effect in the cow.[365] Preparations containing LH arrested luteal regression in heifers hysterectomized on the 10th day after oestrus, and actually caused an increase in the size, and progesterone content, of the corpora lutea. This was done with equine LH, the only LH preparation which, in the earlier experiments already described, failed to exhibit a luteotrophic effect in intact heifers. This is interesting as a hint that the uterus is involved in luteal maintenance (or regression) in the cow. Conversely, oxytocin injections, such as caused luteal regression in the intact animal, failed to do so in the hysterectomized heifer.

Early regression of the corpus luteum in the hysterectomized heifer could be caused by administering an equine antiserum prepared against bovine LH. This antiserum also inhibited progesterone synthesis in luteal tissue slices *in vitro*. Purified bovine LH, on the other hand, stimulated progesterone synthesis in such preparations.

Finally, it was shown that the luteotrophic properties of LH preparations were abolished by incubation with urea. The significance of this observation lies in an earlier demonstration[337] that this procedure inactivates LH without interfering with FSH. The effects demonstrated in the preparations containing LH may, therefore, safely be attributed to the LH itself, regardless of possible contamination with FSH.

The foregoing observations amount to an impressive weight of circumstantial evidence that LH has a major luteotrophic function in the cow. The evidence admittedly falls short of being absolutely conclusive. It has not so far been demonstrated, for instance, that LH secretion falls in the cow during or before the time that the corpus luteum regresses in the normal course of events, and there is no evidence that luteal regression at the end of the normal cycle results from LH withdrawal. Indeed, the same research group provided evidence of a positive luteolytic mechanism, not only in their demonstration that the uterus is involved in luteal regression but also by showing that luteal tissue taken at the 18th day of the cycle had lost its capacity to respond to LH administered *in vitro*. At this stage, even continued high levels of LH failed to halt regression, and this suggests that an irreversible luteolytic process had been set in motion, overcoming the tissue's capacity to respond to luteotrophin. The sheep and the cow are fairly closely related genera, in which the general pattern of the oestrous cycle is very similar. If a clear-cut difference in the underlying mechanism is proved, of such magnitude as to indicate hypothalamic involvement in luteal maintenance in the one species and not in the other, it will further underline the feebleness of the correlation between the phylogenetic position of an animal and its endocrine constitution.

Inherent viability of the corpus luteum

The effects of hypophysectomy and of pituitary stalk section in the pig have been compared[13] in experiments planned along similar lines to those on the sheep, already described. In the pig experiments, however, the corpora lutea regressed prematurely after pituitary stalk section, much as they did after hypophysectomy. This implies that in the pig, unlike the sheep, the pituitary luteotrophic stimulus is under hypothalamic control, as in the cow, and that it is therefore probably LH. This probability is greatly strengthened by the observation, made during the same series of experiments, that the corpora lutea of hysterectomized pigs also regressed fairly rapidly after pituitary stalk section, but could be maintained by the administration of LH, either in extracts of pig pituitary glands or as human chorionic gonadotrophin (HCG).

The corpus luteum of the unmated rat retains its capacity to respond to prolactin for about 80 hours. This was demonstrated by a neat experiment[1] in which ovulation was induced by ovine LH in rats hypophysectomized during pro-oestrus. It was known that in the intact rat the corpora lutea secrete enough progesterone to render the endometrium capable of a deciduomal response for two to three days. The experiment showed that this capacity, and the longer-lasting capacity to respond to prolactin, were both retained in the absence of the pituitary gland.

In brief, therefore, we may say that, among the few species that have so far been critically examined the corpus luteum appears to have a variable but by no means inconsiderable inherent viability, but that in the normal intact animal its function is regulated and its termination controlled by extra-ovarian factors. We must now proceed to examine these control mechanisms, particularly those which 'switch off' luteal activity and bring about the relatively sudden regression of the corpus luteum with impressively accurate timing. These mechanisms are obviously related to those which maintain luteal activity, but they are sufficiently distinct to justify a separate description. That they are distinct is shown by the finding that luteal regression is not caused simply by the exhaustion of the luteotrophic stimulus. The regularity of the cycle argues against this being so, but it remains to be seen to what extent luteal regression is brought about by factors that regulate or antagonize the luteotrophin or by a 'luteolytic' mechanism, possibly an actual humoral agent, a 'luteolysin'.

Uterine control of the corpus luteum; the effects of hysterectomy

We have already alluded to the fact that the corpus luteum of the unmated animal persists beyond its normal period after removal of the uterus in some species but not in others. The effect of hysterectomy on the life-span of the corpus luteum was summarized by Anderson, Bowerman and Melampy in 1963[12] for all species studied up to that time. They cited a large number of authors and although the list of references could be extended, no further species appear to have been examined in this respect since that time. Hysterectomy extends the life of the corpus luteum in the unmated guinea-pig, pig, sheep and cow, but not in the unmated mouse, rat, thirteen-lined ground-squirrel, rabbit, ferret, dog, rhesus monkey or man. The effect is positive during pseudopregnancy (i.e., the persistence of an active corpus luteum is prolonged by hysterectomy) in the rat, mouse, rabbit and hamster. Hysterectomy does not have this effect in the dog or in the ferret, or in the rhesus monkey or man. In the two former species the corpus luteum, once formed, persists in the unmated or pseudopregnant animal for as long as it does in pregnancy, whereas in the menstruating species it regresses after a much shorter time. One is tempted to hope that some clue to the mechanism of the 'hysterectomy effect' is to be found in the very fact that it is exhibited in some species but not in others, but it is difficult to see any relationship, either phylogenetic or physiological, between the groups of species so assorted.

This difficulty may be resolved if the suggestion put forward by one American team is substantiated. They claim to have demonstrated 'a utero-ovarian relationship independent of species'.[78] They had already shown that a subcellular fraction from the guinea-pig endometrium was capable of inhibiting progesterone synthesis by guinea-pig luteal tissue *in vitro*, but only when the enzyme employed was from the guinea-pig ovary. This activity, they found, was restricted to uterine tissue, similar extracts from skeletal muscle or other tissues being found to be ineffective. In their subsequent work they found that fractions from uterine, but not from other tissue, from a variety of species including monkey and man, were effective in these preparations, albeit only about half as effective as those from guinea-pig uteri. The implication is that the uterus, even in species in which hysterectomy does not prolong the survival of the corpus luteum, elaborates something that affects luteal function. Further investigation will doubtless inform us whether this is a finding of wide physiological significance, or whether it applies only in the special circumstances of the in-vitro preparation described. The former case is not inconceivable, since it may be that this uterine potential exists in all species but, in many, is not employed as the principal mechanism for terminating the luteal phase.

Loeb, who described the effects of hysterectomy in the guinea-pig as long ago as 1923,[227] suggested that the uterine endometrium produced an internal secretion, with the function we have termed 'luteo-lytic'. A few years later he drew attention to the similarity between the corpora lutea of pregnancy and what we may conveniently call the 'corpora lutea of hysterectomy'. He suggested that the prolongation of luteal activity in pregnancy was due, at least in part, to the functional inactivation of the uterine mucosa under the influence of the developing embryo and placenta. This work, and Loeb's interpretation of it, seem to have attracted little attention until relatively recently, but in the past few years much research has been directed to this problem, especially in the sheep and pig. This has been due, in part, to the demonstration (see below) of a 'unilateral' uterine effect in these species, and probably in all those where hysterectomy markedly prolongs the per-sistence of the functional corpus luteum. In writing of a 'unilateral' or 'local' effect one is referring to the fact that in some species removal of one horn of the uterus causes the corpora lutea in the corresponding ovary to persist as they would after hysterectomy, while those of th contralateral ovary regress as in the normal cycle. It is possible t explain the 'hysterectomy effect' on the hypothesis that the uterus is site of oestrogen metabolism so that hysterectomy has an 'oestrogen-sparing' effect,[181, 182] the oestrogen being further supposed to exert luteotrophic effect. There is some reason to expect that the uterus shoul metabolize oestrogen since it is, of course, an important target org for this steroid, and there is long-established evidence that oestrog can be luteotrophic in some species.[231, 314] More recently, howev experiments in which oestrogen and/or progesterone were administer to spayed and/or hysterectomized rats have shown that the rat uterus does not utilize measurable amounts of oestrogen but does utilize con-siderable amounts of progesterone.[98] Uterine trauma, or hysterectomy, therefore seemed to have a progesterone-sparing effect, and the pro-longed persistence of the corpus luteum after hysterectomy may be due to the increased amount of progesterone in circulation. The effect of the progesterone, as envisaged on this hypothesis, is twofold: in the first place by positive feedback action on the secretion of prolactin, and in the second place by a negative feedback action on the secretion of gonadotrophin. The resultant inhibition of LH secretion may also be supposed to have a twofold effect. First, it is concerned in the maturation of the follicles that are due to ovulate when luteal regression is sufficiently far advanced and, secondly, there is some evidence that LH is luteolytic in the rat.

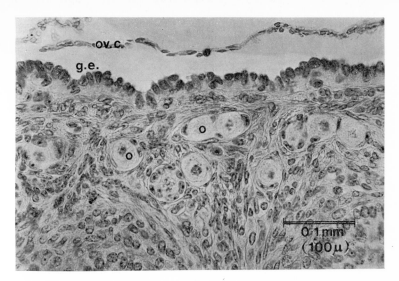

PLATE I. Peripheral region of guinea-pig ovary, (×200) *ov.c.*, ovarian capsule; *o*, primary oocyte; *g.e.*, germinal epithelium.

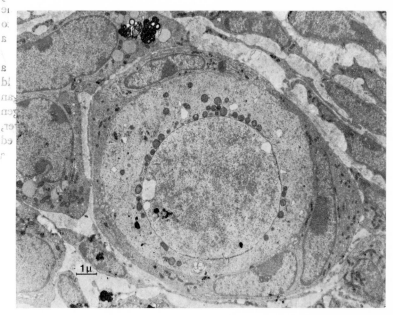

PLATE II. Electron micrograph of a primordial follicle of a guinea-pig. (×5500).

A single layer of follicle cells surround the oocyte. Yolk granules are clustered near the nuclear membrane, forming a cap over one pole of the nucleus. The section is nearly radial to the nucleus, but not to the whole oocyte. The electron-dense mass outside the follicle (top of picture) is caused by autolysis of a dead cell. The stromal cells surrounding the follicle are held in a matrix formed by connective tissue and containing collagen fibres.

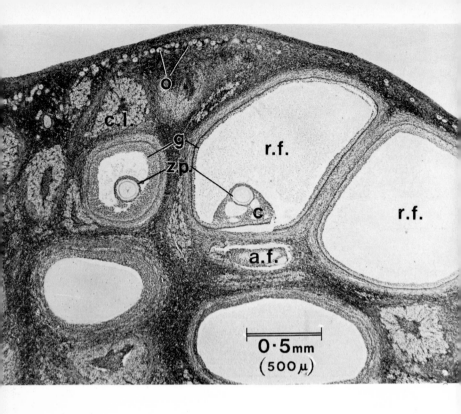

PLATE III. Part of the ovary of a cat, showing ripe follicles and old corpora
lutea. (× 40 approx.).

o, primordial follicles near periphery; *r.f.*, ripe follicles; *a.f.*, an atretic follicle;
g, membrana granulosa; *c*, cumulus oöphorus; *z.p.*, zona pellucida; *c.l.*, regressed corpus luteum

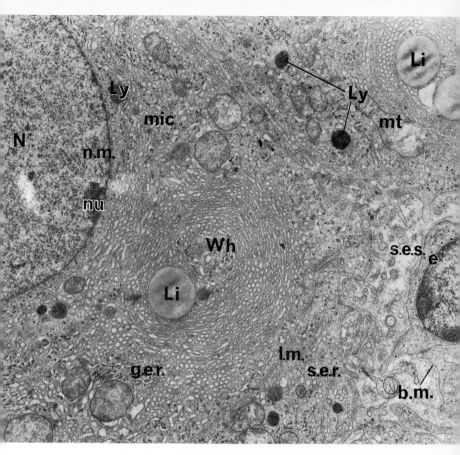

PLATE IV. Electron micrograph of guinea-pig corpus luteum (× 8000).

Part of the nucleus (*N*) of a lutein cell is seen at top left; (*n.m.*) nuclear membrane; (*nu*) a nucleolus. In the cytoplasm of this cell, near the centre of the photograph, the agranular endoplasmic reticulum has the 'whorl' form (*Wh*) characteristic of steroidogenic tissue. It encloses a lipid granule (*Li*). There is other agranular (*s.e.r.*) as well as some granular (*g.e.r.*) endoplasmic reticulum. The border of this cell is produced into numerous tentacle-like processes where it adjoins the subendothelial space (*s.e.s.*). Beyond this 'space' (bottom right) is part of an endoethelial cell (*e*), the basement membrane (*b.m.*) of which is clearly seen. Where two lutein cells lie close together their borders are simple and the limiting membranes (*l.m.*) are smooth. Free microsomes (*mic*) are scattered throughout the cytoplasm of the lutein cells and there are a number of lysosomes (*Ly*). Mitochondria (*mt*) are not numerous.

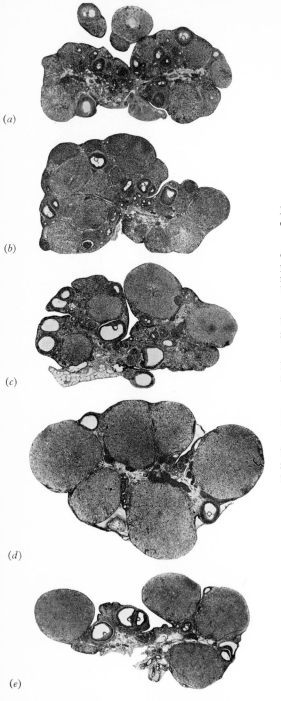

(a)

(b)

(c)

(d)

(e)

PLATE V. The rat ovary at different stages of pregnancy. (×10)

(a) 1½ days after mating; corpora lutea survive from the previous ovulation. One of the new corpora lutea, still hollow, can be seen.

(b) 2 days; new corpora lutea becoming solid; old ones further regressed.

(c) 11 days; old corpora lutea have regressed almost completely; new follicles are growing.

(d) 14 days; corpora lutea have grown; follicles remain in 'resting' condition.

(e) 21 days; corpora lutea have begun to regress; follicles have ripened for the impending postpartum oestrus.

Although the hysterectomy effect may be susceptible to an explanation of this sort, there remains the fact that the corpora lutea do regress in quite a normal way in the hysterectomized pseudopregnant rat. Their regression is merely delayed by a few days, and in these circumstances it is necessarily under other than uterine control. Functional corpora lutea persist in the rat after autotransplantation of the pituitary gland,[125] so that their regression must normally be controlled by extra-ovarian factors, as we have already remarked. LH may have an important role in luteal regression in the rat; it seems probable that this is true of the cow and, perhaps surprisingly, the evidence about this species appears to be more conclusive on this point than that derived from the laboratory rat.

The 'unilateral effect'. Such 'sparing' effects as those postulated in the preceding section appear inadequate to explain the 'unilateral' effect after the removal of one horn of a bicornuate uterus, when the corpora lutea regress in the ovary adjacent to the retained uterine horn but not in the other ovary. The fact that the two members of a pair of ovaries can thus be made to behave quite differently implies the existence of a local route of communication between the uterine horn and the ipsilateral ovary. This immediately suggests a nervous pathway, but it has been shown by experiments involving surgical section of segments of the uterus that such pathways are not involved in this particular phenomenon (but see p. 115). The fact that the prolongation of luteal persistence caused by hemi-hysterectomy is less than that caused by total hysterectomy suggests that the uterine luteolytic influence can accumulate until it is communicated systemically—a systemic effect backing up the initial local effect. Conceivably it may be that the stimulus is not blood-borne at all and the slow effect of a contralateral uterine horn may be due simply to the greater distance between endometrium and ovary.

The 'unilateral effect' was first demonstrated in 1961 by Du Mesnil du Buisson's experiments on pigs.[111] He had already shown that total hysterectomy before the 16th day of the 21-day cycle leads to the persistence of the corpora lutea of the sow for a period at least as long as that of gestation (Fig. 12a). Removal of only one uterine horn did not have this effect (b). After removal of one uterine horn together with all but a short length near the ovarian end of the other, the corpora lutea persisted on the one side and regressed on the other (c). These investigations have been extended both by the French workers and by others, particularly in the United States. There has been fruitful co-operation between the 'Centre Nationale' and the Iowa State University Department of Animal Husbandry, and an interchange of visits between Jouy-en-Josas and Ames, Iowa.

Much of this work has been done on the pregnant sow, including the animal which is caused to have embryos in one horn only (a uni-

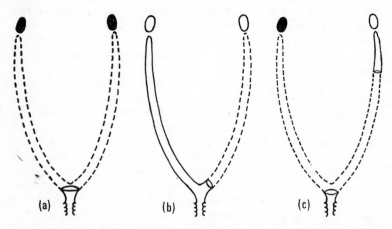

FIG. 12. The 'unilateral effect' in the pig, demonstrated by Du Mesnil du Buisson in 1961. Broken lines represent excised portions of the uterine horns; ovaries with persistent corpora luteau shown black.

lateral pregnancy). This may be achieved simply by ligating the cervical end and transecting the ovarian end of one horn before the animal is mated (Fig. 13a). Embryos rarely survive in such an animal, because there is enough 'non-pregnant' uterine tissue to cause the corpora lutea of both ovaries to regress. They do survive, however, if the 'empty' horn is removed (b) before the 14th day of pregnancy, but not if it is left in place until after the 16th day. The oestrous cycle of the unmated pig is 21 days, and this result is interpreted to mean that the luteolytic influence of the uterus, if not counteracted or inhibited by the presence of embryos, must become effective about the 14th day, and that the corpora lutea must be irreversibly affected by the 16th day.[112] If a short length (about 1/8th) of the 'empty' horn is left in place, and the rest removed, the embryos survive, and the corpora lutea persist in the ovary ipsilateral to the 'pregnant' horn, but those in the other ovary regress. The persisting corpora lutea of the 'ipsilateral' ovary increase their output of progesterone, so partly compensating for the loss of progesterone output from the other ovary.[14]

Removal of the pituitary gland after hysterectomy leads to the rapid regression of the corpora lutea of hysterectomy in the pig.[113] This indicates that their persistence, beyond the duration of the normal cycle, involves a hypophyseal luteotrophic factor. Hypophysectomy during

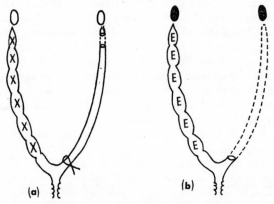

FIG. 13. Unilateral pregnancy in the sow.

(*a*) An 'empty' uterine horn causes regression of the corpora luteau of both ovaries (cf. (*b*) of Fig. 12) and embryos in the contralateral horn die.

(*b*) If the 'empty' horn is removed within the first two weeks of pregnancy, corpora luteau are maintained and embryos can survive in the remaining horn.

the cycle does not cause premature regression of the corpora lutea in this species, and hypophysectomy performed within about 40 hours of ovulation does not prevent the formation of corpora lutea. Their formation and normal life-span therefore seem to be governed by a brief hypophyseal stimulus which is applied early in the process of follicular maturation.

Role of the embryo—transfer experiments. The experiments with unilaterally hysterectomized and unilaterally pregnant pigs very strongly suggest a 'local' influence of the uterine horn upon its adjacent ovary, but speculation beyond this is of doubtful value. It is difficult to account for the corpora lutea being apparently independent of pituitary luteotrophin for the duration of the cycle but subsequently, in hysterectomized animals, acquiring a dependence on it. Perhaps the simplest hypothesis would be that the pituitary luteotrophin (perhaps LH) is released cyclically, in 'bursts' at approximately 3-week intervals. This would provide for luteal survival for one cycle in the presence of an 'empty' uterus. On this hypothesis the uterus is regarded as developing a luteolytic influence sufficient to overcome the viability of the corpus luteum before the next pituitary 'boost'. This luteolytic influence is envisaged as being exerted directly on the corpus luteum, primarily by 'local' pathways, so that resection of one horn and part of the other leaves only enough to effect the ovary adjacent to the remnant. The regression of the corpora lutea of hysterectomy after hypoplysectomy due to the failure of the luteotrophic boost. It is conceivable, perhaps even probable, that LH release is in fact cyclic but of course the real system may not be anything like this model.

Another species now very thoroughly investigated with respect to luteal control is the sheep. A 'unilateral effect' was demonstrated in the ewe by Moor and Rowson in 1966,[257] and this Cambridge team has been chiefly responsible for its further elucidation by a series of surgical experiments. It is pleasant to record the frequent interchange of ideas between the research centres concerned, progress being made by contributions from a variety of differing skills and facilities. The apparent extinction of the uterine luteolytic effect by the presence of a developing embryo, has been studied by the technique of surgical transfer of one or more blastocysts from one ewe to another.[258-260] The oestrous cycle is usually controlled by hormone injections so as to 'synchronize' donor and recipient with respect to the condition of the ovaries and uterus.[324] It was found that if an embryo of equivalent age is transferred to a ewe on or before the 12th day of the cycle, it will survive, and the recipient's corpora lutea will persist into and throughout a normal pregnancy. The presence of the embryo in some way overrides or prevents the lytic effect of the uterus. In the absence of an embryo the corpora lutea persist in an apparently fully active condition until the 15th day. Then progesterone secretion falls abruptly, and only then can any regressive changes be seen histologically, histochemically or electron-microscopically. The presence of an embryo therefore prevents the development of the lytic effect; if an embryo is transferred on the 13th, 14th or 15th day of the cycle, luteal regression is deferred but normal pregnancy is not completed.

Conversely, if the embryo is removed from a pregnant ewe by the 12th day after coitus, luteal regression occurs on the 15th day[259] and the animal 'is never aware that it has been pregnant'.[350] One might say, in fact, that the sheep is not informed about a pregnancy until the day before it is due to ovulate again. Similar effects on luteal maintenance can be obtained by the infusion of homogenates of embryonic tissue obtained from sheep 14–15 days pregnant.[325] Homogenates from 25-day embryos were ineffective so it would appear that the luteotrophic or anti-luteolytic stimulus is only required during a critical period, after which some other trophic stimulus takes over, or else the uterine lytic activity ceases. It will be seen that although patient and elegant experimentation has provided much relevant information we still cannot answer the question which Short[352] attributes to Halban (1904): 'comment le corps jaune sait-il qu'il ya a grossesse?'

In a further series of 'egg-transfer' experiments, involving 115 recipient ewes, the same team transferred 5- or 9-day embryos to various 'isolated' portions of the uterus.[260] The part of the uterus containing the embryo was 'isolated' in that it was severed from the remainder and tied off from it. When an embryo was transferred to an intact uterus, pregnancy ensued and the corpus luteum was maintained, regardless of whether the embryo was in the ipsilateral horn (same side as the corpus luteum) or in the contralateral horn (opposite side). This is illustrated in Fig. 14(a)(b). When one horn of the uterus was 'isolated' and an embryo introduced into it, pregnancy ensued if the embryo and corpus luteum were on the same side (c), but not if they were on opposite sides (d). The embryo and the corpus luteum did survive in the latter case if the empty horn was removed (e).

When an embryo was transferred to an 'isolated' uterine horn in an animal with corpora lutea in both ovaries (f) the embryo survived but the corpus luteum of the contralateral ovary regressed. When an embryo was transferred to the 'isolated' *ovarian* half of a uterine horn, on the same side as an ovary with a corpus luteum (g), the embryo survived in a substantial number of cases; when it survived the corpus luteum was maintained. An embryo placed in the 'isolated' *cervical* half of the uterine horn did not survive, and the corpus luteum regressed even when it was in the ipsilateral ovary (h). Similarly, embryonic death and

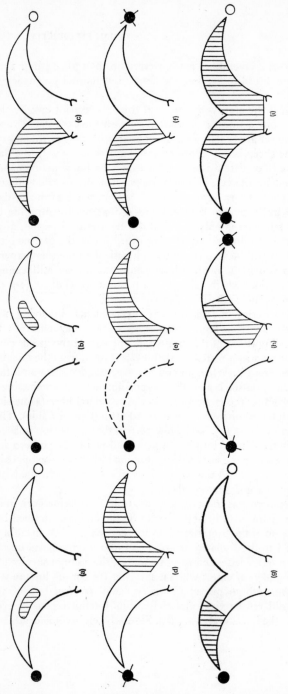

FIG. 14. 'Isolated' conceptuses in the sheep. Experiments by Moor and Rowson, 1968.

Vertical shading represents the position of an embryo, which is in an 'isolated' portion of the uterus in Figs. (c) to (i). A solid circle represents an ovary with a corpus luteum, an open circle represents an ovary without a corpus luteum. Solid circle crossed represents regression of the corpus luteum. See p. 109.

luteal regression followed when the 'isolated' ovarian half of the ipsilateral horn was left empty and an embryo was placed in the other part
of the uterus (i).

Laboratory animals. A 'unilateral' hysterectomy effect on the corpus
luteum has been demonstrated in the guinea-pig. Although total hysterectomy has a very marked effect on survival of the corpora lutea in this
species, as already described, the effects of partial hysterectomy have
not been so clearly demonstrated in it as in the farm animals. In 1967
Fischer carried out various surgical amputations and expressed the
results in terms of the delay of the return of oestrus in the experimental
animals as compared with controls.[131] The main results may be summarized with the aid of a diagram (Fig. 15(a)–(d), in which the broken
lines represent parts that were removed, a solid black circle represents
an ovary with corpora lutea that persisted for a longer time than usual
and an open circle represents an ovary in which the corpora lutea regressed at the normal time.

Total hysterectomy (Fig. 15a) led to prolonged persistence of the
corpora lutea (as in Loeb's experiments, and others, see above) and a very
long delay before oestrus recurred. Hemi-hysterectomy (Fig. 15b) led to
a succession of cycles, irregular but prolonged significantly beyond the
normal length. Both ovaries functioned in a similar manner. Removal of
one uterine horn and the ipsilateral ovary (Fig. 15c) had no effect on
the cycle length, whereas removal of one horn and the contralateral ovary
(Fig. 15d) was followed by prolonged cycles as in (Fig. 15b). These results
appear to show that each uterine horn affects its 'own' ovary more strongly
than that of the other side. This represents strong evidence of a 'unilateral effect' in this species. This evidence is also supported by the results
of experiments in which one uterine horn has been distended by the
insertion of a relatively large bead in the lumen.

In sum, these investigations clearly demonstrate a 'unilateral effect'
in some species and, moreover, they are very difficult to interpret in any
other way than by postulating the influence of a positive luteolytic
activity on the part of the uterus. Even so the evidence for this or for
a local route of communication between uterus and ovary is not universally accepted. As Short expressed it: 'Not only have we failed to
demonstrate how the uterus exerts its lytic effect on the corpus luteum,
we have failed to demonstrate any luteolytic agent. Small wonder,
therefore, that many people are still highly sceptical of this whole
concept'.[354]

Endometrial luteolysin

Several attempts have been made to demonstrate and identify a specific uterine 'luteolysin' by experiments on luteal tissue grown *in vitro*—that is, maintained in a culture medium in an incubator, isolated from other tissues. Pig granulosa cells grown for 7–8 days in this way

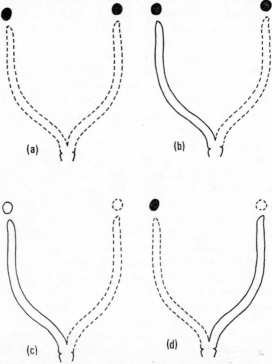

FIG. 15. Fischer's (1967) experiments on partial hysterectomy in the guinea-pig. Solid circles represent ovaries in which corpora lutea presumably persisted beyond the expected time; an open circle represents an ovary in which corpora lutea regressed at the expected time. Broken lines represent excised organs. Hemi-hysterectomy caused persistence of the corpora lutea (cf. pig, Fig. 12) but note that these data are derived from observations of the oestrous cycle length (see text).

were then shown to be secreting relatively large amounts of progesterone.[338] Similar culture medium was flushed through the lumen of
uteri taken from sows immediately after slaughter and the uterine
flushings were added to the cultures after being sterilized. Flushings
from sows in the first ten days after ovulation had no effect but those
from sows later in the cycle especially from those killed 14–18 days after
ovulation caused the cells to shrink and steroidogenesis ceased almost
immediately. These observations provide extremely impressive evidence
that a substance inimical to luteal activity is produced by the uterine
endometrium during a restricted part of the cycle. There are difficulties,
in that the destructive flushings are also deleterious to other cells from
the same animal (e.g., kidney cells) but the effects on granulosa (luteal)
cells were more marked, and pig granulosa cells were more susceptible
than corresponding cells derived from human, sheep or horse ovaries.

The effective agent in these experiments was obtained from the
uterine lumen, and was presumably secreted into the lumen by the
endothelial cells lining it, but it seems most improbable that this is the
route by which it reaches the ovary, should it indeed prove to be the
postulated uterine luteolysin. The 'direct' route up the fallopian tube
to the ovary has been shown almost certainly not to be the normal route,
since ligation of the fallopian tube has no effect on the persistence of
the corpus luteum.

There is still further evidence that luteal regression can be caused
by something elaborated by the endometrium. Moor and Rowson and
their colleagues have recently shown that suspensions of freeze-dried
sheep endometrial tissue cause the breakdown of disaggregated luteal
cells grown in culture on cover glasses. They have not, at the time of
writing, been able to bring about premature regression of the corpus
luteum in the living sheep by infusing such an extract into the ovarian
vein. It is possible that the 'test tube' situation does not accurately
represent the situation within the animal, but the fact that they have so
far been unable to control the corpus luteum in the in-vivo experiments
by no means disproves the validity of their in-vitro experiments. It is
impossible to calculate the rate at which material needs to be infused,
the amount available is strictly limited, the time for which the infusion
can be maintained is also limited, and the animal has to be anaesthetized
for the whole period. The difficulties are obviously immense. On the
other hand a piece of endometrial tissue grafted into the persistent
corpus luteum of a hysterectomized sheep has been shown to cause the

breakdown of the luteal tissue in its neighbourhood. The histological picture is exactly as if some cytolytic substance diffused out of the healthy endometrial tissue and spread through the luteal tissue. If such a substance is responsible for luteal regression in the normal animal it must possess the remarkable property of affecting only 'lutein' cells. This is a degree of specificity that is only to be expected if the lutein cells themselves have a special responsiveness to the lytic substance, a situation very strongly suggestive of a relationship akin to a specific immune reaction.

If this mechanism for controlling the ovarian cycle by means of a uterine luteolysin is eventually proved to exist in some species and not in others, as seems possible or even probable on present evidence, it will represent a divergence between species more fundamental than that between species in which prolactin is luteotrophic and those in which it is not, or even that between menstruating species and those with an oestrous cycle. As the menstruating species, including man, fall among those to which this explanation of luteal regression apparently does not apply, elucidation of the controlling mechanisms of the human cycle will perhaps depend upon the experimental use of other menstruating animals. Laboratory colonies of several primate species have already been established and are slowly increasing in number, but these animals are not yet available in numbers nearly comparable with species such as the sheep and pig. Even these by virtue of their size and cost are unavailable to many researchers, and the rat, mouse, rabbit and guinea-pig still provide the bulk of the material for fundamental work on reproductive physiology.

Experimental uterine distension

We have seen that, in the sheep at least, the embryo appears to exert a direct effect on luteal maintenance. It has long been recognized that the prolongation of luteal activity and the postponement of ovulation is a central feature of pregnancy, as seen from an endocrinological viewpoint. Interesting results have been obtained from experiments in which inert substances have been used to distend the uterus in imitation of the most obvious effect of the developing embryo. Two very different lines of investigation devolve from such experiments. On the one hand they demonstrate that neural pathways play a part in controlling the cycle in the normal animal, and on the other hand they are relevant to the contraceptive effect of the 'intra-uterine device' or 'IUD'.

Let us first examine the former aspect. The oestrous cycle of the ewe was shortened by about four days when plastic beads (8 or 11 mm in diameter) were sutured into the uterus, but this did not happen if the part of the uterus containing the beads was first denervated.[261] Uterine denervation by itself had no effect on the cycle. Subsequent experiments showed that smaller beads (2 mm) had no effect and that the 8 mm beads, which shortened the cycle when inserted on day three, lengthened it when they were inserted on day thirteen.[269] Both the shortening and lengthening effects were prevented by denervation, as in the earlier experiments.

Effects similar to those in the sheep have been reported in the guinea-pig, but were not found in the pig. The mode of placentation is different in all three but the difference between the sheep and the pig is not so extreme as that between these species and the guinea-pig. In any case, as the effects of uterine distension were demonstrated in the sheep at a time in the cycle corresponding to a stage of pregnancy when the embryo does not distend the uterus to this extent, comparison with the mechanics of pregnancy may be unfruitful. Furthermore, the plastic beads affected cycle length at a stage of the cycle when the embryo does not do so, as shown by the experiments of Moor and Rowson already described, and the role of the neural pathways in normal utero-ovarian interaction remains obscure.

The other aspect of 'intra-uterine devices' is their contraceptive action when left in place for long periods. The effect was first put to practical use in the 'Graefenberg ring' more than 30 years ago. As a contraceptive method it fell into disrepute, partly because of infections that were not caused by the ring as such, and it has only recently been revived in different forms usually made of plastic. One disadvantage of the Graefenberg ring was that its insertion required cervical dilatation under anaesthesia; the modern 'coil' or 'bow' does not require this. It is not 100% effective as a contraceptive, although it inhibits pregnancy almost totally. The failure rate is estimated at 26–50 per 1000 woman years, the higher figure including cases due to unnoticed expulsion of the device. It may be more readily adopted when its mode of action is better understood.

There is some evidence that the contraceptive effect of an intra-uterine foreign body is exerted locally rather than systemically, and that it takes the form of a direct toxicity for blastocysts or spermatozoa rather than the inhibition of implantation. Short lengths of material were inserted in the uteri of rats, rabbits and mice. Surgical silk was used in the rats and mice, polythene tubing in rabbits. The rats were of two kinds: the ordinary laboratory animals and 'germ-free' rats, which are specially maintained and fed so as to harbour few or no bacteria. A chronic infiltration of polymorphonuclear leucocytes into the uterine lumen was induced in the immediate region of the foreign bodies in all four types of animal. In the mice and 'conventional' rats an inflammation spread along the uterine horn from the region of the foreign body and the whole length of the uterus was sterile (i.e., infertile). In the rabbits and in the germ-free rats no spread of inflammation occurred, and the parts of the uterus not occupied by the foreign bodies were capable of sustaining embryos. The inflammation appeared not to be associated with the presence of bacteria since none could be cultured from the rabbit uteri or of course from the uteri of the germ-free rats. Lysozyme was found in the uterine lumen of rats and rabbits, suggesting that polymorphonuclear leucocytes released their contents into the lumen. The authors concluded that 'some substance derived from polymorphonuclear leucocytes may exert toxic effects on fertilized ova or on spermatozoa and thus be responsible for the infertility of uteri containing foreign bodies'. Their pictures show polymorphonuclear leucocytes penetrating the uterine mucosa and its epithelium, and show that the inflammation described does not reach acute proportions and does not grossly alter the endometrial histology.[287]

Intra-uterine foreign bodies appear to increase LH production in sheep, and shorten the cycle slightly in guinea-pigs; they have no effect on the cycle of pigs or cows but they depress fertility and interfere with insemination. It may transpire that such devices interfere with cycle length perhaps by nervous pathways in species where the uterus directly controls luteal function and have a quite different effect in other species, such as man. This however is mere speculation.

The ovary and the ovarian steroids in pregnancy

That the primary function of the corpus luteum is to condition the uterus in preparation for implantation of the embryo, and to maintain pregnancy, has been known for many years. Earlier work in this field was reviewed by Amoroso and Finn in 1962,[8] and a more recent article by Short[355] provides a summary of what is known of the sequence of endocrinological events in several of the species in which most progress has been made. We have already alluded to the fundamental modification of the ovarian cycle that this function involves, the delay of ovulation and the persistence of the corpus luteum; this is universally characteristic of the eutherian mammals, but the manner in which it is achieved, and the changes in ovarian activity during the course of pregnancy, vary enormously within the order.

Pre-implantation changes in the uterine mucosa are much more striking in some species than in others, but although they may not be obvious they are essential. Implantation depends on a critical balance between progesterone and oestrogen and, after implantation, maintenance of the conceptus still requires circulating progesterone. This hormone may be supplied entirely by the corpora lutea (pig) or the luteal supply may be augmented by placental progesterone (guinea-pig) or, as in man and in the mare, the corpus luteum may regress in midpregnancy. In such species the placenta provides the only major source of progesterone for the remainder of the gestation period.

The stage of gestation at which the corpus luteum can be dispensed with is readily estimated by ovariectomy. This causes abortion or resorption of the embryos when carried out at any stage of pregnancy in the pig, whereas in an animal such as the horse, the ovaries of which contain no corpora lutea after about mid-pregnancy, ovariectomy after this time does not interfere with gestation. In man the average gestation period is 267 days, and ovariectomy has been performed as early as the 40th day without terminating pregnancy. The ovary is not necessarily inactive when it no longer secretes progesterone, since the interstitial tissue may secrete other steroids. In animals in which the corpus luteum persists throughout pregnancy it is also involved in the control of parturition and of mammary development.

Implantation

When the blastocyst reaches the uterus, it lies free in the lumen—for 4 days in the mouse, guinea-pig and rabbit, for example, and longer in some species, such as the macaque, and the cat; implantation occurs about 10 days after ovulation in the macaque, and about 14 days after it in the cat. In all these species the egg takes about 3 days to reach the uterus. This period seems to be very general among mammals but there is evidence that tubal passage is accomplished in about 2 days in the pig. The ferret is a more striking exception, for in this species the egg spends about 6 days in the fallopian tube and lies free in the uterine lumen for a further 7 days.

In the case of polytocous species one might suppose that some time would be required to distribute the eggs within the uterine cavity, but this would not apply to monotocous species, nor is it known by what means blastocysts are moved, spaced and positioned. The ovary is involved because both the mobility of the uterine myometrium and the reactivity of the endometrium are controlled by ovarian steroids. There is no evidence that intra-uterine movement of eggs or blastocysts has any relation to the length of time they spend free in the uterine lumen.

The interval between the egg's entry into the uterus and its implantation in, or attachment to, the uterine endometrium is very much longer in some species, often occupying more than half the total time between ovulation and parturition. There are, in addition, species such as the rat, in which the free uterine stage is significantly lengthened in pregnant animals which are concurrently suckling a large litter. The term 'delayed implantation' is usually applied both to the very long intervals that occur normally in some species and to those which are lengthened because implantation is literally 'delayed' beyond the normal by a recognizable cause. The former are conveniently referred to as having an 'obligate' diapause, as distinct from the 'facultative' diapause of the second group.

The species in which these phenomena have been investigated were reviewed by Deanesly in 1966 in the course of a comprehensive essay on the endocrinology of pregnancy,[95] and *Delayed Implantation* forms the subject and the title of a symposium volume edited by A. C. Enders in 1963.[118] Deanesly listed 18 eutherian species in which the delay is of the order of weeks or months. They include members of the Artiodactyla, Edentata and Carnivora. The phenomenon is a remarkably frequent characteristic of mustelid species and of the seals, and it also occurs in some bears. The earliest record relates to the roe deer, in which delayed implantation was discovered by Ziegler in 1843 and by Bischoff in 1854. It has recently been suggested that the initial discovery should be credited to the physician William Harvey, in 1651, but 'Close inspection of Harvey's original work in Latin makes it abundantly clear that he was only studying the red deer (*Cervus elephus*) and the fallow deer (*Dama dama*), two species in which delayed implantation is not known to occur.'[356] Australian biologists have described delayed implantation as a regular feature of the reproductive cycle of several marsupial species (see p. 6). In *Setonyx* and *Protemnodon*, for example, the single blastocyst derived from a post-partum mating remains unimplanted while the young of the previous pregnancy is in the pouch. If the pouch young is removed, the embryo implants.

The rabbit, like the mouse, rat and some other rodents, may become pregnant at an immediately post-partum oestrus. This is rare in laboratory rabbits, which breed throughout the year, but occurs regularly during the restricted breeding season in wild rabbits.[40] In the rabbit, however, implantation is not delayed in a pregnancy that originates at a post-partum oestrus and accompanies lactation. In the ferret, which, like the rabbit, has 'induced' ovulation (see p. 78), pregnancy rarely accompanies lactation, but such evidence as exists suggests that when it does, implantation occurs at the usual time. The field vole, *Microtus agrestis*, has recently been shown to be another species in which ovulation is induced by copulation, at least in a well established laboratory colony, and implantation is only slightly delayed during a pregnancy that is concurrent with lactation.[44] The suggestion that the period of delay in implantation may be related to a pause in the growth of the corpus luteum appears to be attributable to Brambell.[39] It was based on data for the shrew (*Sorex*) collected in the wild state, and actual measurement of the time intervals was impossible, but the evidence suggested that the corpora lutea ceased to grow just before the eggs entered the uterus, and that their growth was not resumed until the time of implantation or immediately afterwards. The corpora lutea even appeared to regress a little in the meantime. In the same paper Brambell directed attention to the work of Lataste who, in 1887, recognized the relation between lactation and the delay of implantation.[216] Brambell suggested that there might be a significant similarity between the inhibition of oestrus during lactation (lactation anoestrus) and the prolongation of pregnancy due to delayed implantation; he referred to earlier work showing that the inhibitory effect of lactation on oestrus is exerted through the ovaries.[284] He discounted the possibility that the delay could be a direct effect of the nutritional demands of concurrent lactation and pregnancy, because it may occur when relatively few young are being suckled (5 in rat, 3 in mouse) and in any case one could not imagine that the embryos, being minute at this stage, could constitute a significant drain on the maternal organism. He further remarked on the fact that the corpora lutea resumed growth after, rather than before, implantation, and suggested the possibility that the blastocysts are inhibited by some substance secreted into the uterine lumen. Since that time a large body of information has been amassed concerning the endocrinology of implantation, but the process is still incompletely understood.

Many have seen in the phenomenon of delayed implantation a potentially useful experimental system for the investigation of the normal event. There is no doubt that changes in the ovarian hormone output are responsible for the delay, and for the resumption of activity, but the nature of their effects is still obscure. Deanesly, in reviewing the evidence, showed that both oestrogen and progesterone are involved, and that the effectiveness of one of these hormones in facilitating implantation, in a situation where it would otherwise be delayed, may be attributed to its synergistic action with the other. Generally speaking, it would seem that sufficient progesterone is available for implantation, which therefore awaits additional oestrogen. However, the amount and the timing of the oestrogen requirement lie within critical limits, so that it may be easier to accelerate implantation with additional progesterone. That this can be done was neatly demonstrated by Meyer and his colleagues, who actually obtained superfoetation in rats that were suckling large litters. They injected progesterone directly into the uterine musculature of one side just before the time when ova, fertilized at a post-partum oestrus, were expected to enter the uterine horns. Embryos developing in the progesterone-treated horn implanted with no more than the normal (non-suckling) delay, whereas those in the other horn did not implant until some days later.

Among the species in which the blastocysts lie free in the uterine lumen for weeks or months, the embryonic diapause may or may not be associated with lactation. In some, such as the seals, it would appear to have an obvious adaptive value in extending the gestation period so that the animal can give birth to one litter, and conceive the next, during a single brief sojourn ashore. A similar argument could also be applied to other species, such as the stoat, in which implantation is controlled by daylength, with the result that both mating and parturition occur in early spring, the mating season being spring or early summer. The influence of daylength and of other 'external factors', on the ovarian cycle is discussed elsewhere (p. 164), but at this point it is relevant to mention Hammond's demonstration that implantation in the mink can be brought forward by exposure to light but not by the administration of progesterone.[161] From the fact that the luteal cells enlarge after a sterile ovulation just as they do before implantation, Hansson had earlier argued that light affects implantation in mink through an ovarian effect. After removal of the ovaries, implantation occurred normally when progesterone was given by twice-daily injections for a week.[166] Thus Hansson showed that the ovarian factor involved was in fact progesterone. Hammond concurred in this, but by showing that progesterone alone did not cause earlier implantation, he showed that the delay or diapause, as such, was not simply due to a progesterone deficiency.

Brambell's suggestion as to a direct effect on the developing egg, exerted through the luminal fluid, is recalled by the observation made 30 years later that mouse blastocysts retain the zona pellucida for longer in lactating females when implantation is delayed than in 'normal' pregnancy.[238] This was in line with the observation that when mice are ovariectomized early in pregnancy the ova do not implant and loss of the zona is delayed.[278] The mouse eggs did not retain the zona pellucida until implantation, but the loss of the zona pellucida seems to play a very significant part in implantation in the rabbit.[146] When rabbit endometrium was maintained in organ culture, and eggs were allowed to implant on it, they did so under a variety of experimental conditions, but it was always necessary to remove the zona pellucida beforehand. Another indication of an interaction between the embryo and the medium in which it lies before implantation was noted in the course of work on early pregnancy in the pig. Eggs (in the morula stage) that were apparently abnormal or dead retained the zona pellucida when it had been lost from living embryos in the same litter.[294]

There appear to be no data relating to blood progesterone levels during pregnancy in animals with delayed implantation. The growth of the corpora lutea has been described in several, and its progesterone content has been measured in some. The roe deer is unusual in that the corpus luteum is apparently fully active throughout the period of delay, as judged by its weight, histological appearance and progesterone content.[356] The information now available concerning the probable role of 'protein binding' in the control of progesterone metabolism (see p. 133) suggests that a mechanism of this kind should be sought in the case of the roe deer, but this does not imply that the resumption of embryonic

E

activity is not initiated by a hypothalamic response to some environmental stimulus.

Some recent work in France was designed to investigate the role of progesterone in relation to early embryonic survival in the sheep.[392] It had already been shown that a large proportion of embryonic loss is incurred early in pregnancy in the sheep as in a number of other species. The French team embarked on a project designed to define the critical period more precisely and, at the same time, to study the importance of progesterone in early embryonic survival. By superovulation, alone or in combination with surgery, they were able to study survival among several embryos in the presence of several corpora lutea, and among single embryos in ewes with several corpora lutea. In the former situation the acceleration in blastocyst growth, from the 8th day p.c., lasted only until day 10 or 11, whereas it continued for a longer period where there were several corpora lutea per embryo. There appeared to be a relation between the rate of embryonic growth and the amount of progesterone secreted 'per embryo'; but this was true only within restricted limits, beyond which the further addition of progesterone had a deleterious effect. The authors found that the amount of progesterone (actually the amount of steroidogenic tissue in the ovary) was of little significance for the first 7 days after mating. After this, they concluded, a balance is established between the number of embryos present and the total amount of progesterone available. Embryonic growth accelerates and, at the same time, the embryo develops an increased sensitivity to high levels of progesterone. Beyond an optimal level, progesterone becomes detrimental for this reason, and the embryos that are the first to be stimulated by progesterone are the ones that first succumb to the supraoptimal quantity.

Gestation and pro-gestation

Although the process of implantation varies so greatly among the orders, and even genera, of mammals, it always constitutes a critical stage in the establishment of definitive placental relations between embryo and uterus. In animals such as the rat and the guinea-pig, the blastocyst can be seen to be lying free in the uterine lumen at one moment, and a few hours later is found to have 'burrowed' through the uterine epithelium and sunk into the uterine tissue. This process, and the somewhat similar attachment of the still-minute blastocyst in man, clearly resemble the planting of a seed, and the descriptive term 'implantation' is obviously appropriate to them. The word 'nidation' is a synonym, apparently derived from French usage. Many authors now distinguish the pre- and post-implantation stages of pregnancy as 'progestation' and 'gestation' respectively This usage is convenient within the context where it is usually applied, but confusion is possible because 'gestation' is a well established synonym for 'pregnancy' (as in the familiar 'gestation period' meaning the lapse of time from fertilization to parturition). It has become clear that the early luteal phase is similar in pregnant and non-pregnant animals, whether mated or not, so that the restriction of 'gestation' to the period of established pregnancy is all the more reasonable. One would, however, be hard put to it to define the moment of implantation, or the change from progestation, or the change from progestation to gestation, in the pig.

Because of its well defined nature in the rodents, as well as their other advantages as experimental animals, it is in them that implantation has been most intensively studied. Particular attention has centred around the role of the two principal ovarian steroids, and the balance between them.

The oestrogen 'surge'. The synergism of small amounts of oestrogen with less critical amounts of progesterone has been demonstrated in rats in which implantation normally occurred on day 6 (if a vaginal plug was found on day 1).[95, 253] When the ovaries were removed on day 4 and progesterone was then administered daily, implantation occurred at the usual time, provided a small amount of oestradiol was administered on day 4. When the oestrogen was injected later, implantation was delayed. This embryonic diapause could be prolonged, under continued progesterone treatment, for several weeks, and a suitable small injection of oestrogen at any time during this period would be followed by implantation within 48 hours. From these results, taken by themselves, one might be tempted to suppose that the survival of blastocysts was ensured by progesterone, while their implantation was caused by oestrogen. Further experiments, however, showed that the blastocysts survived in the uterus of rats ovariectomized on day 4 and given no steroid replacement. Such blastocysts could be reactivated, and caused to implant, by the administration of 10 mg progesterone with 1 μg oestrogen. It was thought that the survival of blastocysts in ovariectomized rats might be due to small amounts of progesterone secreted by the adrenals, but they survived even when the adrenals as well as the ovaries were removed. Their mere survival, therefore, appears to be independent of progesterone. However, progesterone is perhaps beneficial, and is certainly not harmful, to the blastocysts, whereas oestrogen administered in the absence of progesterone was found to have a severely deleterious effect on them.

If the ovaries were not removed until day 5, implantation proceeded normally under progesterone treatment, without oestrogen. From this observation, and other experiments, it is evident that the rat ovary secretes a critical amount of oestrogen during a restricted period of time on day 4. It may be significant that this oestrogen 'surge' (as it has become widely known) occurs at approximately the time when oestrogen output would be increased in the non-pregnant rat, but it is thought that the mechanism involved in pregnancy is, in fact, different from that of the oestrous cycle. The sequence of events is certainly modified, of course, for oestrous and ovulation are suppressed.

Autotransplantation of the pituitary gland (that is, its removal to another place in the same individual) maintains it as an organ but severs its connection with the hypothalamus. Like pituitary stalk section (p. 99) this results in the continued secretion of prolactin, and the cessation of gonadotrophin secretion. Because prolactin is luteotrophic in the rat, the corpora lutea are maintained, but follicles do not mature and produce oestrogen. When this operation is done within a day or so after copulation, it results in delayed implantation. This seems to imply that the lack of oestrogen in such animals is due to the inhibition of follicular growth—the corpora lutea produce progesterone, but the follicles fail to produce sufficient oestrogen. It must be remembered, however, that even the first few days of pregnancy in the rat are different from the corresponding days in the oestrous cycle, for the corpora lutea are functional in pregnancy and not in the unmated animal. Follicular growth is suppressed in pregnancy, so that even in the intact animal there are no ripe follicles to account for the oestrogen surge, and many investigators consider that this oestrogen, as well as the necessary progesterone, is produced by the corpora lutea.[346] If this is so, the ovary's failure to produce oestrogen after autotransplantation of the pituitary implies that FSH or LH, or both, are required to maintain oestrogen production by the corpus luteum, although its histology and, apparently, its ability to produce progesterone, are maintained by prolactin alone.

The guinea-pig is like the rat in that implantation is sudden and 'interstitial', and the blastocyst 'burrows' into the endometrium in much the same way. Deciduomata can be produced by traumatization of the endometrium during the luteal phase, which in the guinea-pig is part of the normal oestrous cycle. Whereas in the rat, however, it is possible to induce deciduomata in this way while blastocysts lie free in the uterine lumen, guinea-pig blastocysts are able to implant even in a uterus that is resistant to this treatment. As one writer put it: 'The blastocyst showed itself a more efficient agent for inducing decidualization than a needle and thread.'[94] Traumatization is more effective in producing deciduomata if additional progesterone is administered and it has recently been found that it is particularly effective in the lactating guinea-pig.

In the rabbit, implantation appears to depend on progesterone alone, for it occurs in ovariectomized animals if this hormone is given. The process is perhaps facilitated by oestrogen, however, for deciduomata cannot be induced by endometrial trauma in the ovariectomized rabbit unless both progesterone and oestrogen are administered. The decidual reaction is less immediate in the rabbit than in the rat or guinea-pig; this is to be expected since the mode of implantation is 'central'. The blastocyst does not sink into the uterine stroma (as in the rat and guinea-pig) but becomes distended to a sphere about 3 mm in diameter before the trophoblast begins to 'attack' the uterine epithelium.

We have seen that rat embryos (blastocysts) can survive but not implant in the absence of progesterone, and it has also been shown that guinea-pig embryos can implant and begin to develop without progesterone—at least without the ovarian contribution.[94] Placental progesterone is sufficient to enable embryos to survive after ovariectomy from about day 20 (in a pregnancy of about 66 days), and ovarian progesterone is absolutely necessary only during the six or seven days preceding this. Before this, however, embryonic growth is decidedly retarded, and no embryos survive beyond day 16 in untreated ovariectomized animals. In hypophysectomized guinea-pigs, as in ovariectomized ones, blastocysts can implant, but fail soon afterwards.

In the European badger, in which there is normally a long embryonic diapause or 'delay', implantation cannot be induced prematurely by the administration of progesterone. The reproductive physiology of this interesting animal was described by Neal and Harrison in Britain,[271] and it has also been investigated by Canivenc and his collaborators in Bordeaux.[62] The French authors wrote of their 'great difficulties in obtaining the animals needed', and in the same sentence mentioned the 600 badgers they had studied up to that time! The species is widespread, and the breeding season appears to vary in different geographical regions. It was shown, however, that in some regions at least, the young are born in February and a fertile mating occurs immediately after parturition; the zygotes develop to the blastocyst stage and then lie dormant in the uterine lumen for 10 months. If the ovaries are removed during the long embryonic diapause, the blastocysts survive for many months, but they fail to implant. On the other hand, the administration of ovarian hormones does not hasten implantation.

At the time when the blastocysts of the badger suddenly begin to grow, reaching a diameter of 5 mm preparatory to implantation, the corpora lutea undergo a marked histological change. Each corpus luteum increases in size, evidently by the hypertrophy of the individual luteal cells. There is little progesterone in the 'latent' corpora lutea—that is, during 'delay'—and a very much higher content in the active and enlarged corpora lutea after implantation.[63] Although it is not possible to advance implantation by administering progesterone to the badger, it is reasonable to conclude that the change in the corpora lutea is causally related to the rapid growth and implantation of the blastocysts. Increased amounts of progesterone are evidently necessary at this time and subsequently, but they are of no use earlier. It therefore seems probable that the increased progesterone secretion is accompanied by a change in metabolism, so that the hormone is not only available, but effective. A binding protein may be involved, such as that to which progesterone is bound during pregnancy in the guinea-pig (see p. 133).

One may presume that whatever stimulus brings about implantation in the badger acts through the anterior pituitary, and therefore through the hypothalamus. The badger is nocturnal and fossorial, and the critical period involved falls in December. The French investigators suggested that, because of its strictly nocturnal habit, the badger would be most active when the days are shortest, and they postulated that this factor governs the ovarian changes. They inserted a small radio transmitter beneath the skin of a badger and recorded its activity by radiotelemetry for almost a year. This confirmed their impression that the period of greatest activity was in December, and they wrote: 'We think this is not a mere coincidence. We believe that in the badger the prolonging of the nocturnal phase conditions the luteal activity in the same way as prolongation of the length of day stimulates ovarian activity in the mink or the marten'. This, however, is in marked contrast with Harrison's statement about badgers in Britain. He says: 'In the wild, activity diminishes during December; badgers spend less time above ground and sleep for long periods in a chamber filled with hay or other heat-insulating material'.[173] The difference in habits is conceivably attributable to the difference in climate between the localities concerned, but it is difficult to imagine that implantation would depend on increased activity in one locality and on decreased activity in another. Both French and English authors are agreed, however, that light seems unlikely to be the decisive factor. An exchange of animals would perhaps be interesting, but this species does not command economic priority, and only limited resources have hitherto been available for such projects.

Has the study of delayed implantation provided much information, as promised, about the actual mechanisms involved in nidation? It has certainly provided some and, as might be expected, the most profitable approach, so far, has been to exploit the control over timing made possible by the facultative delay experimentally induced in such species as the rat. At the same time, the recurring problem of 'species differences' restricts the applicability of such results. Furthermore, progress to date has been limited to the increasingly precise description of hormonal requirements, with little information about the mode of action of the hormones involved.

Ectopic implantation and uterine delay

A very interesting experiment, bearing on the cause of delay in the absence of oestrogen, was described by the late David Kirby in 1967.[213] He removed the ovaries of a number of mated mice while the eggs were still in the fallopian tubes, and then administered progesterone by daily injections. In these conditions, implantation was delayed, but the blastocysts remained viable. Some of them were flushed from the uterine horns after the normal time of implantation; others were left in the uterine lumen. Those that were flushed out were transplanted to the mother's kidney capsule (autografts). It was shown that the transplanted blastocysts 'implanted' and developed more or less normally for some time. Those that were left in the uterine horn did not implant unless a suitable dose of oestrogen was administered.

In describing these results, Kirby refrained from speculating about the role of oestrogen or the reasons why ectopic implantation occurred without it. In fact, the experiment was an extension of earlier work in which he showed that blastocysts develop more readily outside the uterus than within it, once the initial intra-uterine stages have been accomplished.

Homografts can be made to a variety of locations in animals of either sex, and in some cases mouse embryos can even be grown in male rats. Indeed, Kirby has shown that far from being the embryo's only safe refuge the uterus is, rather, the only place where it is safe to turn a blastocyst loose. So it would seem that the oestrogen acts on the uterus to relax its resistance; the blastocyst itself needs no further stimulus. Mayer, in the review already quoted, referred to the effect of oestrogen in rats as being 'to bring the blastocyst out of its lethargy', but unless the situation is very different in this species, this seems to misplace the hormone's target in the initiation of implantation.

To revert to the badger: the change in luteal metabolism associated with implantation may involve the production of oestrogen, but attempts to induce implantation in badgers by administering oestrogen, either alone or in combination with progesterone, have not been successful. Attempts have also been made to advance implantation in a badger by removing the ovaries early in diapause. When the animal was killed, some time after the operation, the blastocysts had not implanted but they were found to be larger than those of other animals killed at a corresponding time.[173] 'Lack of live animals prohibited further experiments' and the problem apparently rests at present. The basis of this experiment on a badger was presumably the earlier discovery that, in the armadillo, ovariectomy in the middle of the 4-month embryonic diapause is followed by implantation about a month later, the embryos dying shortly afterwards.[54] The authors thought that this result indicated that 'some non-ovarian tissue assumes the function of maintaining the uterus', but in subsequent experiments they were unable to bring about the same result by hormone administration, and wrote that 'The enigma of implantation following bilateral ovariectomy in this species remains unexplained'. The phenomenon is compatible with the idea that it is normal for the uterus to resist the trophoblast's invasion, but this requires that the ovary is responsible for this restraint as well as for the subsequent conditioning that permits embryonic development after implantation.

Accessory corpora lutea in pregnancy

In a number of species, widely scattered among the various orders of mammals, several follicles form corpora lutea besides the ones from which the fertilized ova are derived. Such 'accessory' corpora lutea are often formed by the luteinization of follicles that do not ovulate, but in some species they are regularly formed from follicles that do so. The bizarre case of the African insectivore *Elephantulus myurus*[378, 380] where as many as 60 ova may be shed from each ovary but implantations are restricted to two, has been known for nearly 30 years. Very recently an even more extreme multiplicity of corpora lutea has been described in the plains viscacha (*Lagostomus maximum*) of Argentina. As many as 1,000 ova may be shed at oestrus and all the ruptured follicles, and many others besides, are converted into corpora lutea.[387]

Accessory corpora lutea are also prominent in pregnancy in the mountain viscacha (*Lagidium peruanum* Meyen)[289] and in the Canadian porcupine (*Erithizon dorsatum*),[265] both of which, like the plains viscacha, are hystricomorph rodents. In the former species they are formed from the 40th day onwards, in a pregnancy of about 100 days, and as many as 12 accumulate before term. In the Canadian porcupine accessory corpora lutea are formed in large numbers in both ovaries at oestrus and in early pregnancy by the luteinization of atretic follicles. They persist, however, only in the ovary which contains the corpus luteum of conception. In the nilgai, a big Indian antelope (*Boselephas*), the ovary is packed with corpora lutea, all similar in size, during pregnancy[7] and the same is true of the collared peccary.[393] This, like the pig and the hippopotamus, is a non-ruminant artiodactyl.

Another exceptional but very different sequence of events occurs in pregnancy in the horse. Here, the 'corpus luteum of conception' begins to regress about a month after ovulation, and a number of smaller corpora lutea are formed by the ovulation of successive follicles, one or more at a time, over a period of two to three months[74, 318]. These corpora lutea secrete very substantial amounts of progesterone, but they all regress by about 150 days after ovulation, and during the remaining 200 days of pregnancy the ovary of the mare has a fibrous and 'inactive' appearance. The ovary of the African elephant, and the difficulty of interpreting events within it on the evidence available, has already been referred to. In this species, the ovaries of pregnant animals sometimes contain only one corpus luteum, but in the pregnant specimens I collected there were usually several corpora lutea in each ovary.[291] It was suggested that a number of follicles ovulate more or less simultaneously, while others perhaps luteinize without ovulating and that, usually, only one of the ova is fertilized, or only one fertilized ovum implants (twins are rare). It was further suggested that the luteal apparatus is renewed about mid-pregnancy by the formation of a new set of corpora lutea that lasts to term. Similar material, in much greater quantity, is now becoming available to biologists in East Africa, and a re-appraisal of this problem may be expected.

The mink is unique among the species so far described in that 'accessory' ovulations may add to the embryos in the litter during pregnancy. This is possible because delayed implantation, or embryonic diapause, is a normal feature of pregnancy in the mink and the supernumerary ova are produced during this period. The mink ovulates in response to copulation, so the accessory ovulations must be accompanied by oestrus; the 'original' blastocysts are small and are lying free in the uterine lumen and so do not prevent spermatozoa from reaching the additional ova. At the end of the diapause the additional blastocysts become implanted together with the original ones.[119, 166] Hansson recorded two cases of superfoetation in the mink, presumably resulting from the implantation of 'accessory' blastocysts at a different time from that of the orginal ones; this must occur very rarely. Accessory corpora lutea are also formed from ovulatory follicles in the badger, another species in which delayed implantation is common. Oestrus and copulation do not, apparently, occur at the time of the 'accessory' ovulations and the ova are not fertilized.

Relatively insignificant accessory corpora lutea are formed, sporadically, in the ovaries of many species during pregnancy, including man. They are common in the wild Norway rat, but relatively rare in the albino laboratory strains of the same species. They probably have little physiological significance.

'Luteinization' of the foetal ovaries is characteristic of some species at certain stages of pregnancy. In the giraffe a number of follicles develop and 'luteinize'. In the horse and the elephant the interstitial tissue of the foetal ovary or testis is greatly hypertrophied and the ovaries of a foetal horse may be larger than those of its mother. As the source of PMSG, the mare deserves a section to itself, and this phenomenon is considered in more detail elsewhere (p. 138).

Growth of the corpus luteum and the secretion of progesterone

Besides persisting for a longer time than those of the unmated cycle, the corpora lutea of pregnancy often grow to a larger size. In the guinea-pig, for example, the corpora lutea of the cycle reach their maximum size, approximately 2·5 g, in about 8 days.[179, 319] During pregnancy the corpora lutea grow at a similar rate at first, but they continue to grow for nearly twice as long and attain a weight of up to 5 g. They remain about this size until very near the end of pregnancy and then decline abruptly. The additional growth in pregnancy, beyond that of the cycle, begins soon after implantation and appears to be related to it—the time relations suggest that the additional luteal growth depends on implantation having been achieved, rather than implantation depending on a luteal stimulus. The uterus has been 'aware of' pregnancy, or at least prepared for it, for some days; actual invasion of the decidua by the trophoblast, and elongation of the 'embryonic cylinder' occurs about the 7th day after mating. The corpora lutea of the pregnant rat also attain about twice the weight or volume of the corpora lutea of the unmated cycle but, as shown by Fig. 9, they do not grow at a steady rate. After the first week of pregnancy, they cease to grow, or even regress a little in size, increase a little about day 11, and then suddenly double their volume. They begin to regress an appreciable time before the end of the relatively short gestation period. The sudden increase in size nearly two-thirds of the way through pregnancy is difficult to relate to any embryological event or endocrinological crisis. The definitive allanto-chorionic placenta is well established by this time, and the level of circulating progesterone has been falling since day 10.[150]

Techniques for the estimation of progesterone in small amounts of tissue or of body fluid have only been developed within the past few years, but plasma progesterone levels at different stages of pregnancy have been measured in a number of species. In some, such as the goat and sheep, the concentration does not rise significantly above that of the luteal phase of the unmated ('pseudopregnant') cycle, although it is, of course, maintained for a longer time. In others, such as the guinea-pig and rat, and in human pregnancy near term, the circulating levels are many times higher than in the unmated animal.

In experiments using guinea-pigs,[179] it was found that, in this species, hypophysectomy early in pregnancy did not, in most cases, prevent embryos from implanting, and the corpora lutea survived and grew as in normal pregnancy. Sixteen of 27 animals killed up to four weeks after pre-implantation hypophysectomy contained normal conceptuses, the corpora lutea were normal in size and appearance, and the plasma progesterone levels were normal. In others, only regressing embryos were found at autopsy; apparently implantation took place but the embryos failed after this. In some, there were no embryos at all (showing that implantation had not occurred) and in these animals the corpora lutea had regressed by the time of autopsy. When embryos were found, whether they were normal or regressing (i.e., in all cases when implantation had occurred) the corpora lutea were of normal size, but only in animals with some healthy embryos was the level of plasma progesterone within the normal range for pregnancy. The continued function of the corpus luteum was, therefore, apparently related to the occurrence of implantation, as the time relations had already suggested.

The question then arose as to the functional status of the corpus luteum that 'looks' normal, histologically, and contains progesterone, in an animal that has a low level of circulating progesterone in the blood. The level in the plasma of non-pregnant guinea-pigs is low—not measurably different from that of ovariectomized animals—and, surprisingly, the level in hysterectomized guinea-pigs was also found to be low, although the corpora lutea resembled those of pregnancy in size, appearance and progesterone content. The question was whether a low plasma progesterone concentration in an animal with a high luteal progesterone concentration implied slow production, prolonged accumulation in the luteal tissue and slow release into the circulation, or normal production and release (secretion) with rapid metabolism and elimination from the bloodstream. The latter alternative was suggested by a neat experiment in which tablets of progesterone were inserted beneath the skin of ovariectomized guinea-pigs, a technique known to provide the hormonal requirements for normal pregnancy. The rate of absorption of progesterone from the tablets was the same whether the animal was pregnant or not, but plasma progesterone levels were high in the pregnant and low in the non-pregnant ones, whether the uterus was removed or not.[177] This experiment makes it appear very probable that the plasma progesterone level depends on the rate at which the steroid is eliminated from the plasma, and not on the amount released from the ovary. The authors further suggested that their findings could well be related to the observation that transcortin, a corticosteroid-binding globulin, has a high affinity for progesterone; it is even higher, at body temperature, than its affinity for cortisol.[142, 332, 342] There is a marked rise in the amount of circulating transcortin during pregnancy in the guinea-pig, at a time that roughly corresponds with the rise in plasma progesterone but follows rather than precedes it. In an earlier paper, Heap and Deanesly had shown that the circulating progesterone level is maintained by the placenta in the later stages of pregnancy; they found no significant difference in the plasma progesterone level between intact and ovariectomized guinea-pigs in the last two weeks of pregnancy.[176] On the other hand, the concentration of progesterone in the corpora lutea remains as high during this time as at earlier stages[322] so it would seem that, during this period at least, luteal content and output do not correspond. This has to be borne in mind in evaluating the evidence from the tablet-bearing animals, but it appears extremely probable that changes in the transcortin level, or in that of a specific progesterone-binding protein, more recently discovered,[175] constitute the operative factor in the control of plasma progesterone levels in the guinea-pig. This latter protein, so far known only in the guinea-pig, is believed to make its appearance rather earlier than the rise in transcortin.

Further evidence pointing to this conclusion is provided by measurements of the metabolic clearance rate (MCR) of progesterone at different stages of pregnancy. The MCR is high throughout the cycle in the non-pregnant guinea-pig and during the first two weeks of pregnancy. It falls dramatically in the third week of pregnancy.[203] The MCR is calculated as the volume of blood cleared in unit time. It depends on the concentration and is therefore not directly related to the rate of metabolism. It is reasoned, however, that a high clearance rate of progesterone is causally related to a low concentration of the steroid in the blood, and that the clearance rate falls, and the plasma concentration consequently rises, when the binding protein appears in the circulation. The method employed to measure the MCR of progesterone was the slow infusion of a small amount of radio-isotopically labelled (tritiated) progesterone into an anaesthetized guinea-pig, by way of the jugular vein, until (after about 3 hours) there was no further rise in the radioactivity of progesterone isolated from blood samples withdrawn at intervals from the carotid artery. This implies that the labelled steroid has reached equilibrium throughout the tissues, and the clearance rate of 'radio-activity' may be assumed to balance the input. As the concentration of progesterone (labelled and unlabelled) in the plasma is measured at the same time, the rate of its elimination can be calculated as the volume of blood cleared per unit time per unit body weight. Progesterone was found to be very rapidly eliminated from the blood in mid-cycle guinea-pigs (MCR: over 100 l/day/kg). During the third week of pregnancy the MCR suddenly dropped to less than 10 l/day/kg.

Rowlands and Heap measured luteal size, luteal tissue progesterone and plasma progesterone levels in the coypu (nutria, *Myocastor coypus*) and compared the data with those for the guinea-pig.[321] The comparison is interesting, since both species are members of the rodent sub-order Hystricomorpha, a group which is characterized by long gestation periods. Pregnancy in the guinea-pig lasts 65–70 days; this is a long gestation period for an animal of its size, but it is short compared with many other hystricomorphs. In the chinchilla, which is smaller than the guinea-pig, the gestation period is 111 days. It is 104 days in the agouti and 112 days in the African porcupine. In the largest member of the group, the capybara, which is the largest rodent, pregnancy lasts 120–126 days. The female coypu may reach a body weight of 8 kg or even more, but weighs 1·5–2·0 kg at sexual maturity and the mean gestation period is about 130 days.[276] The long gestation period is associated with the birth of the young in a very advanced state of development in most hystricomorphs. Comparable information concerning the ovarian changes is not yet available for many of them, but what is known suggests that considerable variety may be expected within the group. In the guinea-pig, as we have seen, the corpus luteum grows to its maximum size in the first third of pregnancy and retains this volume virtually to term. The plasma progesterone level rises in similar fashion, to maintain a concentration of about 200 ng/ml. In the coypu the corpora lutea of conception are usually supplemented by accessory corpora lutea. Their number ranged from 0 to 15 in Rowlands and Heap's material, most animals having 2 to 6. It was not possible to distinguish the corpora lutea of conception from the accessory corpora lutea and it is therefore possible, though unlikely, that the surplus corpora lutea, in excess of the number of embryos, resulted from embryonic loss. The relatively insignificant excess of corpora lutea over embryos in the guinea-pig is probably the result of embryonic loss, but the data for the viscacha and for the Canadian porcupine, referred to earlier, together with preliminary reports of investigations on other hystricomorphs, suggest that the occurrence of accessory corpora lutea may be widespread within the sub-order.

In the coypu, the corpora lutea grow slowly for about the first third of pregnancy, to about 2 mm³, then begin to grow more rapidly and reach a mean diameter of about 9 mm³ by the 100th day. They soon begin to regress, and they shrink very rapidly during the last 30 days of pregnancy. Plasma progesterone levels begin to rise before the sudden acceleration in luteal growth, but their peak approximately coincides with that of the mean volume of luteal tissue, around 100 days. The progesterone concentration of the arterial plasma at this time reaches the extraordinarily high level of about 500 ng/ml. It then declines rapidly as the corpora lutea regress.

We have already described the evidence that the placenta augments, and later supplants, the ovary as a source of circulating progesterone in pregnancy in the guinea-pig. The work of Rowlands and Heap strongly indicated that this does not occur in the coypu, and they further suggested that the hormone level was maintained by the corpora lutea under the influence of a luteotrophic stimulus, perhaps from the placenta. In their words: 'Firstly, ovariectomy of the pregnant coypu was rapidly followed by abortion. The results of this experiment contrast with those in the guinea-pig in late pregnancy and suggest that the coypu is dependent on the ovarian supply of progesterone for the maintenance of pregnancy. Secondly, the sudden increase in the level of systemic progesterone in the second half of pregnancy was probably not of placental origin since values in uterine vein blood were consistently less than those in systemic blood, whereas in the guinea-pig they were usually much higher . . . Significant increases in the size of luteal cells, in size and weight of corpora lutea and in the amount of progesterone in corpora lutea, all coincide closely with the period when progesterone reached its highest value in systemic blood. The evidence suggests, therefore, that the increased level of blood progesterone found in the second half of pregnancy in the coypu originates from luteal rather than from placental tissue, and that it is effected by a stimulus to the luteal cells, possibly placental in origin, which influences their growth and secretory activity, rather than from a change in the metabolism, excretion or utilization of the hormone. It is clear, therefore, that the pattern of progesterone secretion differs considerably in these two closely related species'. It is, of course, difficult to assess the degree of phylogenetic relationship between two species; the differences between the coypu and the guinea-pig are at least of generic magnitude, but they are similar in many respects.

It is interesting to compare the rabbit and the ferret with respect to pregnancy maintenance. They belong to different orders (Lagomorpha and Carnivora respectively) but share the important characteristic that, in both, ovulation is dependent on coitus. They are also similar in that the ovaries are necessary to the maintenance of pregnancy at all stages, and the corpora lutea appear to be the sole major source of progesterone. In both, the corpora lutea behave very similarly whether the animal is pregnant or pseudopregnant (as by mating with a sterile male) except that pseudopregnancy lasts as long as pregnancy in the ferret and not in the rabbit. The progesterone concentration in ovarian venous blood and peripheral arterial blood in the rabbit rises steadily to a peak about mid-pregnancy and then steadily declines.[256] It follows much the same course in pregnancy, and also in pseudopregnancy, in the ferret, but corresponding data for pseudopregnancy in the rabbit are apparently not available.

The fact that arterial progesterone concentrations are similar in pregnancy and pseudopregnancy in the ferret strongly suggests that the placenta does not produce this hormone. This corroborates the histochemical investigation by Galil and Deane, who found no evidence of progesterone synthesis in the placenta of the ferret at any stage of gestation.[144] In this investigation, steroidogenic tissue was identified by the following method: an unfixed tissue section is incubated in a medium containing a suitable substrate, the co-enzyme diphosphopyridine nucleotide (DPN) and a tetrazolium salt. The tetrazolium is reduced to formazan in the presence of Δ^5-3β-hydroxysteroid dehydrogenase. This enzyme is known to hold a key position in the synthesis of steroid hormones (see p. 32), and its presence or absence in a tissue appears to be diagnostic of this activity. The appearance of formazan, which is opaque and distinctive under the microscope, indicates the presence of Δ^5-3β-hydroxysteroid dehydrogenase in the tissue.

Galil and Deane directed attention to several outstanding problems in making comparisons of this kind. One such difficulty arises from the fact that Galil had already shown that the uterine venous blood of pregnant ovariectomized ferrets contained progesterone.[143] In the light of the histochemical work, it seems probable that this progesterone was of foetal rather than of placental origin; it may, for instance, have been formed by the foetal adrenal glands.

The comparison between the ferret and the rabbit, with regard to ovarian activity, is limited by the great difference in the activity of the ovarian interstitial tissue in the two species. This tissue is very conspicuous in the rabbit ovary in gestation, especially in the latter half of pregnancy, and its steroidogenic activity at this time was demonstrated in a survey of a number of different organs in several species.[326] The authors commented on the apparently low steroidogenic activity of the corpora lutea at this stage of pregnancy, the more remarkable because it has been shown that the corpora lutea as such, and not merely the ovaries, are necessary to pregnancy in this species at this time. They suggested that the corpora lutea must either: (a) produce progesterone by some synthetic pathway other than that by which it is produced elsewhere in the same animal at the same time, or (b) produce, in the latter half of pregnancy, some other substance which is also essential to the maintenance of gestation. They further suggested that, in the latter case, the essential product may be relaxin. This hormone had been shown to synergise with oestradiol and progesterone in the maintenance of pregnancy in the spayed mouse,[155] and they directed attention to the need for further work on this intriguing problem, but it still awaits investigation. Dr Helen W. Deane died in 1964, after a long illness, and this field of research lost a vigorous worker and a delightful person.

In the goat, sheep, pig and cow, plasma progesterone concentrations rise to levels somewhat in excess of those in the luteal phase of the 'normal' cycle, and remain near this level until the last stages of pregnancy. Even allowing for the fact that the ovaries of the sheep and goat usually contain one or two corpora lutea, those of the pig about a dozen and those of the cow only one, ovarian activity is not quite similar in all these species. In the goat, as in the pig, but not in the sheep, the ovary is essential to the maintenance of pregnancy at all stages and it may be presumed that the corpus luteum is the main source of progesterone throughout gestation. Linzell and Heap found that during late pregnancy the ovaries of anaesthetised goats produced progesterone at a rate of up to 10 mg/day, and those of sheep only about 2 mg/day. The goat placenta produced no progesterone, that of the sheep up to 14 mg/day.[225] In a goat with a corpus luteum in only one ovary, the concentration of progesterone in the venous blood leaving that ovary was several hundred times higher than in the blood leaving the other ovary.[178] This is as one might expect, but it had earlier been reported that, in the goat, the progesterone levels were about the same in the ovary containing a corpus luteum and in the opposite ovary containing a large follicle.[308] The more recent investigation, however, also produced evidence of another progestagen in ovarian venous blood, and the implication would seem to be that methods of chemical estimation of steroids had improved in the intervening decade, and that the earlier technique failed to distinguish between progesterone and a related steroid. It is established that other steroids may have progestational properties (hence the general term 'progestagen', by analogy with 'oestrogen') and that they may be produced by the ovary, but other naturally occurring steroids have not been found to have a degree of activity comparable with that of progesterone, and they have not been shown to contribute significantly to the maintenance of pregnancy, except in the rabbit, as already described.

In the pig[251] and in the cow[301] plasma progesterone levels fall near the end of pregnancy, in advance of luteal regression, but the situation is not like that described (above) in the guinea-pig, because in these species the ovaries are necessary to the maintenance of the later stages of pregnancy. In the pig, ovariectomy at any time in pregnancy leads to abortion, and although ovarian progesterone secretion falls by about half between 12 and 2 days before term, there is evidently a sufficient amount from this source alone. In the cow, the ovaries are not, apparently, necessary to the last stages of gestation. At least, pregnancy has been maintained in some cows after the corpora lutea were squeezed out of the ovaries. It is unlikely that significant amounts were produced by the displaced corpora lutea and it is probable that the placenta of the cow contributes to the circulating progesterone at least in the last stages of pregnancy. If this is so, the steroidogenic activity of the placenta is in descending order from sheep to cow to goat among the three domestic species with syndesmochorial cotyledonary placentae. The 'superficial' epitheliochorial placenta of the pig is very different in structure.

The ovaries of the mare in pregnancy. Ovarian relations during pregnancy in the mare are unique in many respects. The species does not, therefore, fit neatly into any ordered scheme, but rather demands a section to itself. It is, of course, the only one among the Perissodactyla about which much is known. For although rhinos are killed in large numbers in East Africa, their ovarian cycle appears not to have been studied yet.

In the mare, as we have seen, the corpus luteum of conception regresses after a few weeks, and a number of accessory corpora lutea are then formed, and these last longer but have also disappeared before mid-pregnancy. Short measured blood progesterone levels in the mare throughout the course of pregnancy, and showed that the concentration is comparable with that of other domestic ungulates during the first three to four months, but then falls dramatically to levels below the sensitivity of the then available methods of estimation—certainly to virtually insignificant levels.[349] This is in startling contrast to the situation in other species, where the plasma level of progesterone is maintained (by the placenta) if the ovaries cease to secrete it. Later, however, Short found that the placenta of the mare is in fact a comparatively rich source of progesterone.[350] He suggested that pregnancy must be maintained, after about the half-way stage, by a local effect of this placental progesterone. It may be metabolized in the placenta and so never enter the maternal circulation or, more probably, it may enter the bloodstream as the free steroid, not bound to a protein, and therefore be very rapidly metabolized, mainly by the liver. Such a local activity of progesterone may be made possible by the diffuse epitheliochorial type of placenta in the mare. The only other species with a 'diffuse' placenta, among those for which we have relevant information, is the pig; as we have seen, the pig's placenta does not produce progesterone, its corpora lutea remain functional to the end of pregnancy, and its ovaries behave very differently from those of the mare.

Pregnant mare's serum gonadotrophin (PMSG). That the pregnant mare's serum contained a potent gonadotrophin (now known as PMSG) was discovered by Cole and Hart in 1930,[72] and the value of the source became apparent after 1935, when a satisfactory method of extraction was discovered. A number of investigations followed.[7] The hormone was found to be concentrated in cup-shaped depressions in the endometrium, and to be present in greater concentrations in the serum of mares of small breeds of horse than in large ones. It was shown to be associated with the formation of the accessory corpora lutea.[9] Even so, much remained to be learned about the unique sequence of events in this species. This was largely because horses are not easily available for study in most laboratories, and partly because they have virtually ceased to be regarded as 'agricultural' animals, at least in Britain. The horse, however, is far from being an animal of no economic importance, even in this country, for the British bloodstock industry has a large annual turnover, and an exceptional stallion may be valued at well over a quarter of a million pounds. Furthermore, reproductive performance is obviously an important factor in this industry, and fertility is notoriously low in pedigree mares. Stud fees vary greatly but they may be as high as £5,000 for a single service; nearly half the services fail to result in pregnancy, but the fee is payable whether the mare 'holds' or not. These considerations were behind some recent investigations in Cambridge, in which the morphology and histology of the endometrial cups, and their gonadotrophin content, were related to ovarian events and serum levels of progesterone.

The FSH and LH activities of PMSG reside in or on the same protein molecule, and the development of a suitable anti-serum to this protein, and thereby a sensitive and relatively rapid immuno-assay for PMSG, was an important step forward.[5] It was confirmed that the rise in gonadotrophin levels coincided with, and probably accounted for, the 'luteinization' of follicles to produce the accessory corpora lutea already described. At this point, it may be mentioned that, whereas these follicles were formerly thought to ovulate, it is now suspected that they do not. The evidence for the ovulation of these follicles was the recovery of unfertilized eggs from the Fallopian tubes at the appropriate time.[9] Some recent observations, however, suggest that such eggs may often be recovered from the Fallopian tubes of adult mares, because unfertilized eggs, in this species, do not pass through into the uterus but remain for quite long periods in the oviducts, and accumulate there. How the tube can distinguish between fertilized and unfertilized eggs is as yet unexplained.

Plasma progesterone levels reach an early peak, with the development of the large corpus luteum of conception (the 'primary' corpus luteum of pregnancy). The progesterone level begins to fall after about 20 days. PMSG production starts soon after this, and accessory corpora lutea begin to form before the progesterone level has fallen by more than about 30%. With the formation of the accessory corpora lutea, the progesterone level rises again, and between the 40th and 80th days of pregnancy it reaches a second peak comparable with the first peak. This is also the period of maximum production of PMSG which afterwards declines rather sharply. By about 150 days p.c. the accessory corpora lutea have disappeared, and circulating levels of progesterone and of PMSG approach zero.

Two curious facts emerged from this work: there is an astonishing tenfold variation between individual mares in the peak level of PMSG produced, and the amount produced tends to become less in successive pregnancies in the same mare. The latter observation suggests that its production is somehow related to a sort of immune reaction between mother and foetus. The wide range of individual variation may have a similar explanation; at least it is conceivable that it is related to the degree of mutual incompatibility between mother and foetus, though such ideas lie in the remoter realms of speculation at present. It is not clear to what extent, if at all, the variation in PMSG production is related to fertility. It may be that this curious system originated by 'making use of' a by-product or of something which individuals may synthesize more of than they require.

The immunological method of measuring PMSG was also applied to the donkey, and to the two interspecific crosses. A mare, pregnant to a male donkey (jack) produces a mule; a jenny (female donkey) pregnant to a stallion, produces a hinny. In the normal pregnant donkey (jack × jenny) PMSG levels are very much lower than those in the normal pregnant mare, but in a jenny carrying a hinney the PMSG levels are even higher than in the mare. In a mare carrying a mule the levels are even lower than in the normal donkey pregnancy. The size of the foetus, relative to its mother, is evidently an important factor in determining PMSG production.

The endometrial cups form as ulcer-like depressions in the mucosa; their histology has been clearly described.[6] Their hormonal product is removed by the bloodstream until about 60 days p.c., so that they are truly endocrine glands up to this time. After this, when the overlying epithelium is broken down and the decidual tissue within the cup starts to lose its histological integrity, the hormone is released into the uterine lumen. Very recently, it has been suggested that this 'decidual' tissue is in reality of foetal (trophoblastic) origin.

The distribution of the endometrial cups in the uterus is not random. They form only in a ring in the lower segment of the pregnant horn, but this location is not determined by a local specialization of the endometrium in this area. Instead, it appears that the cups form in the annular zone of the uterine wall that corresponds to a thickened band of chorion at the limit of the allantochorion. At this stage of pregnancy the horse embryo lies in a nearly spherical chorionic sac. The 'upper' hemisphere of this sac is occupied by the allantois, which wraps around the embryo in its amnion. The 'lower' hemisphere is still occupied by the yolk-sac. The allantois is growing and the yolk-sac is shrinking, but at this stage they are nearly equal in size, so the limit of the allantois, in its contact with the chorion, is roughly equatorial. This condition, and the annular thickening of the chorion, were described and figured (as the 'trophoblastic girdle') in 1897,[127] but it was not clearly associated with the formation of endometrial cups.

Yet another peculiarity of the horse is the great increase in size of the foetal gonads in the latter half of pregnancy. It occurs in both sexes, and is caused by a remarkable degree of hypertrophy of the interstitial tissue. During the same period, large amounts of oestrogen are excreted in the urine of the mare. H. H. Cole and his collaborators, in 1933, suggested that this oestrogen was secreted by the hypertrophied foetal gonads. Later, however, they revised this opinion, for 'Although there is an interesting correlation between the development of the fetal gonads and the concentration of oestrin in the urine (the maximum concentration of oestrin in the urine of the mare occurs at the time when the fetal gonads reach their greatest development and the subsequent fall in the oestrin content of the urine [occurs] at the time when the fetal gonads are regressing), the evidence . . . does not support such a theory'.[64] All foetal tissues studied contained about the same concentration of 'oestrin' (as the oestrus-producing hormone was first called; this term has since been superseded by the general term 'oestrogen') so its presence in the gonads appeared to be merely incidental, and it lost 'the special significance which we had previously attached to it'. Because of this, and the very sudden fall in oestrogen after parturition, Catchpole and Cole concluded that this oestrogen was placental in origin. Very soon afterwards, experiments involving ovariectomy in the pregnant mare confirmed this view.[174]

The foetal gonads of the horse grow from a length of 2–4 mm and a weight of 100–200 mg during the gonadotrophic phase of pregnancy, to a length of 6–7 cm and weight of about 150 g in the eighth month, in the oestrogenic phase, when they are twice the weight of the mother's ovaries.[10] Similar changes have been shown to result from the action of oestrogen on foetal or immature gonads in other species (p. 130).

We have already referred to the fact that, although the placenta makes progesterone during the latter part of pregnancy in the mare, and evidently takes over pregnancy maintenance from the ovaries, this hormone is not released into the circulation, but appears to be metabolized at or near its site of production. It is natural to suppose, therefore, that it is converted into oestrogen, which is released in the bloodstream and eliminated in the urine, incidentally causing hypertrophy of the foetal gonads in its passage through them. I have not, however, read of any experimental work to confirm this. Oestrogen is also excreted in the urine in later stages of pregnancy in man and in other primate species.

Luteal volume in relation to litter size

In pregnant women, plasma progesterone levels soon exceed those of the menstrual cycle, and they rise at an increasing rate which is markedly accelerated in the last few weeks before term.[350] The concentration is much higher in women bearing twins or triplets than in those with a single foetus. Since the corpus luteum regresses after about six weeks it is to be presumed that the source of progesterone throughout most of the gestation period is the placenta, and that the higher values found when there are twin or triple placentae in the uterus are related to the greater volume of tissue producing it. It is doubtful whether an increase in plasma progesterone concentration, above a certain threshold or possibly optimal level, is of any advantage. In a polytocous animal (rabbit, pig or rat) bearing a large litter, the number of corpora lutea can be reduced somewhat without apparent prejudice to the litter, but when the amount of luteal tissue is reduced below a certain level, the amount of progesterone produced may limit the number of foetuses that can be supported. Another aspect of the quantitative relation between luteal volume and embryonic viability has been demonstrated in the rabbit.[2] In this species, egg transport through the fallopian tubes, and consequently implantation, requires an amount of progesterone sufficient to overcome the effects of the oestrogen which is still dominant during the first day of pregnancy. Larger breeds, with correspondingly large ovaries, numerous follicles and consequent high oestrogen output, require a large number of corpora lutea during these initial stages. Rabbits weighing about 3·5 kg needed at least eight corpora lutea at this stage.

The ovary at parturition

Relaxin

There is some relaxation of the pelvic ligaments at parturition in many mammals. Among the more familiar species, this phenomenon is most clearly exhibited by the guinea-pig. One can easily pass a finger between the right and left pubic bones of a guinea-pig during the last few days of pregnancy, and the sacro-iliac joints also relax in a similar fashion. The latter symphysis is more responsive than the pubic symphysis in some other species. The softening of the rigid frame around the passage which the foetus must take at birth is brought about by the action of a polypeptide hormone, appropriately known as *relaxin*. The impressive degree to which the effect is carried in the guinea-pig is presumably related to the large size of the foetus relative to that of the mother. In the mole, whose hindquarters are especially narrow, the pubic bones fuse above the uro-genital tract, which is thus excluded from the pelvic girdle.

There seems to be no doubt that relaxin is produced by the ovary, and in some species at least it is also produced by the placenta, and 'It appears that the ovaries may be the principal source of relaxin in animals (such as sow, rat and mouse) in which this organ is needed throughout gestation, but that the placenta may produce relaxin in those species in which gestation can continue to term independently of the ovaries'.[156] The guinea-pig is included in the latter group, and pelvic relaxation occurs normally after hypophysectomy during pregnancy, provided that the placentae remain attached and functional. The hormone has not been successfully isolated, and tissue extracts are assayed for relaxin-like activity chiefly by their effect on the guinea-pig pubic symphysis; measurements are made in 'guinea-pig units' (GPU).

It is clear, however, that this elusive hormone has other effects as well as pelvic relaxation; among them is an effect on the myometrium: 'It has been established that (relaxin) is capable of inhibiting spontaneous uterine motility *in vivo* and *in vitro* and ... of inducing spontaneous delivery of living young in prepubertal and adult pregnant mice spayed and maintained on progesterone.[8] It has also been reported that the administration of relaxin reduced the amount of progesterone required to maintain pregnancy in the spayed mouse, but the validity of this conclusion is doubtful because the mice were 'primed' with oestrogen and there is evidence that oestrogen itself has this effect of reducing the amount of progesterone required. The pelvic effect of relaxin is only exerted when oestrogen is present in the circulation.

Relaxin has been described as 'a hormone of pregnancy' in reference to the fact that it is not found in non-pregnant females or in males.[95] A number of investigators have described its effect on enzymic and metabolic changes in the uterus, and its importance throughout the greater part of gestation, and not merely towards the end of pregnancy.

In the rhesus monkey, for example, the pronounced hypertrophy and hyperplasia of the endothelium of certain vessels of the uterine endometrium is probably controlled by relaxin produced by endometrial granulocytes (that is, within the uterus itself) in response to circulating progesterone.[195] The sequence of events seems to be that progesterone acts upon oestrogen-primed endometrium to promote its growth and increase the vascular area. Relaxin is required to bring about the changes, which are very striking, in the endothelial cells. These changes, described as 'endothelioid cytomorphosis' by Wislocki and Streeter in 1938, are only observed between the 17th and 29th days of a pregnancy that lasts about 160 days.[394] It has therefore been suggested that they are in some way related to early placental development.

Progesterone

Although relaxin plays a conspicuous role in facilitating parturition, there is no evidence that it is instrumental in initiating it. Some 40 years ago, Allen and Corner showed that parturition could be delayed or prevented, in rabbits, by the administration of progesterone just before the end of pregnancy.[4] Since that time the role of progesterone in the initiation of parturition has been studied extensively. The concept of a 'progesterone block' to myometrial activity, or to its stimulation by oxytocin has been put forward as an explanation of the quiescence of the uterine muscle during gestation.[83, 84] Progesterone is therefore seen as the primary regulator of uterine muscular activity and the principal factor in the initiation of parturition. The uterine muscle is, throughout pregnancy, intrinsically capable of performing the mechanical work that it is actually called on to perform at term. According to the hypothesis, this effort is called into play at the appropriate time by a combination of factors principal among which is the 'withdrawal' of progesterone.

The guinea-pig differs from the rabbit in that pregnancy cannot be prolonged by administering progesterone systemically, and progesterone appears not to inhibit the response of the guinea-pig myometrium to oxytocin, as it does in the rabbit. It is possible, however, that the difference is related to the fact that the placenta, and not the ovary, is the principal source of progesterone in the later stages of pregnancy in the guinea-pig. It is conceivable that because of this the local concentration of progesterone within the uterine wall is relatively very high.[305]

It is striking that, in species where the corpus luteum remains large and active in the later stages of gestation, it regresses suddenly just before term. It is therefore natural to suppose that this ovarian change controls the termination of pregnancy, but it has become evident that this is an over-simplification, if not a complete misinterpretation. Among other complexities, the importance of protein binding in the turnover and availability of progesterone has to be considered. In the guinea-pig, for example, progesterone is eliminated from the blood at a much slower rate in the latter two-thirds of pregnancy than at other times. The lower metabolic clearance rate (MCR) appears to be associated with a high progesterone function, but recent work in our laboratory has shown that parturition is not preceded or accompanied by a rise in the MCR.

The role of the foetus

The idea that the foetus itself must have some part in 'deciding' the time of its quitting the womb has been with us from ancient time. Thus Hippocrates wrote that 'When the child is grown big . . . he incontinently passes out into the outside world'. In relatively recent times experimental investigation has tended to suggest that the timing of parturition, as well as the energy required for its accomplishment, depends on uterine and placental function rather than the stage of development reached by the foetus. However, it has long been known that anencephalic human foetuses are associated with prolonged pregnancy, and certain teratological conditions in the domestic ungulates, involving the pituitary gland, have also been found to lead to prolonged gestation. In 1967, Liggins and his collaborators showed that when the pituitary gland of the foetal sheep was destroyed late in gestation, parturition failed to occur at the normal time, if at all.[222] Later, the phenomenon was more closely defined when it was shown that the result of foetal pituitary ablation was due to its effect on the foetal adrenal glands. The conclusions were summarized very succinctly by Liggins in the following words: 'These results show that the foetal lamb . . . is capable of actively initiating parturition by a mechanism which involves the pituitary-adrenal system. However, it was not established that this mechanism is responsible for the initiation of parturition at term in the normal ewe. Nevertheless, there is some evidence that this may be so. First, it had been shown previously that inhibition of pituitary-adrenal function caused failure of initiation of parturition at term . . . Secondly, the rapid growth of the foetal adrenal which follows the administration of ACTH mimics the pattern which occurs in normal lambs at term.'[221] Earlier work had shown that the foetal adrenal gland of the sheep approximately doubles in weight in the last week of gestation.[75] Liggins added that, although the mode of action of cortisol (the main cortico-steroid secreted by the foetal lamb adrenal) in this connection is unexplained, 'An attractive possibility, since it brings together various hypotheses, would be that (it) interferes with the 'progesterone block' of the myometrium'.

In the light of work on the protein binding of progesterone— referred to above—it is also attractive to suppose that the foetal cortisol may compete with progesterone for attachment to binding protein(s) within the placenta, thus causing drastic changes in progesterone metabolism, especially in the immediate vicinity. However this may be, it is evident that the role of progesterone in the initiation of parturition (or perhaps one should say; in long opposing and then suddenly permitting it) will not be fully explained until its mode of action at the cellular level is better understood. It seems clear that it is 'free' progesterone that exerts a hormonal effect, but is susceptible to very rapid elimination from the circulation. 'Bound' progesterone may provide a 'reservoir'— if the free and bound compartments are in equilibrium, and if the free is instantly replenished from the bound when the equilibrium is disturbed, then minute quantities of free progesterone in circulation will suffice for great hormonal activity, provided the supply is 'buffered' by sufficient amounts of the bound steroid, as in the guinea-pig where progesterone activity seems to be largely controlled by changes in the circulating amount of binding protein. Such changes have not been demonstrated in the human or in the sheep, but this does not necessarily rule out the effect of a 'competitive' effect within the conceptus, especially as Liggins showed that, whereas cortisol administered to the foetal lamb advanced the time of parturition, cortisol administered to the ewe did not.

Oestrogen

Oestrogen seems to have a supporting role wherever progesterone plays a leading part. It evidently has a direct effect on myometrial activity, and a variety of subsidiary but nonetheless important effects throughout the uterine wall, besides its synergism with progesterone.

In many species the principal ovarian activity towards the end of pregnancy is the growth of follicles in preparation for ovulation immediately after parturition. This must presumably release a plentiful supply of oestrogen into circulation and when, as in the guinea-pig, oestrus follows within hours of parturition, this high oestrogen concentration must coincide with the last day or so of pregnancy. This is in marked contrast to the situation in man, or in the domestic ungulates. In the latter, the lapse of time between parturition and the next oestrus is a matter of great commercial interest.

5: External Factors Affecting Reproductive Activity

Puberty, and the initiation of cyclical activity

The mammal reaches sexual maturity by a gradual process. In a sense, this process may be said to begin when the initial differentiation of the gonad determines the sex of the individual, but it accelerates, to a more or less marked degree, over a variable period which precedes the establishment of regular cyclic activity and full fertility. The term 'puberty' which describes this period, is useful although it is admittedly vague. It derives from human experience, and is defined in Butterworth's Medical Dictionary[239] as 'The epoch in a person's life at which the sex glands become active'. An 'epoch' may be 'a period of time', or 'an event from which subsequent events are dated' (O.E.D.). In fact, the term 'puberty' is used in both senses: in relation to adolescent experience it covers a protracted period, but in relation to a full human lifespan it is an event marking the beginning of sexual maturity.

The hypothalamus of new-born rabbits and rats, and 1- to 2-month-old calves, contains relatively large amounts of LH–RF. This activity may well exert a trophic effect on the developing pituitary—'perhaps in the control of gonadotrophic cellular differentiation of the anterior pituitary.'[61] Earlier work had shown that the gonadotrophin content of the anterior pituitary rises gradually during the period between birth and puberty, and from this point of view puberty is a gradual and quantitative, rather than sudden and qualitative, endocrinological event.

Puberty in man

Puberty in girls is associated with the first menstruation (menarche). Full sexual maturity, regular menstrual cycles and normal fertility are not established until some time later. The intervening period is recognized as one of 'adolescent sterility' but this must not be taken to imply that conception never occurs at the first ovulation.

The tendency to earlier maturity among the present generation of children, compared with their parents, has recently attracted much attention. The available data were reviewed by Tanner in 1967.[371] He pointed out the shortcomings of some of the data, especially those based on the age at menarche 'recollected' by women who were questioned as long as twenty years afterwards. Even so, 'The main conclusion is clear. The data are impressively consistent . . . Evidently menarche in Europe has been getting earlier during the last 100 years by 3–4 months per decade'. The trend is observed, and is remarkably similar, in a wide variety of countries and environments. 'Good' data are available for women in Manchester where, in 1820, menarche occurred among 'working women' at an average age of 15·7 years, and among 'educated ladies' at 14·6 years. There is little evidence that the present trend towards earlier maturity has reached its limit, and Tanner predicted that it will probably continue for at least another decade or two. However, it is clear that puberty cannot have been getting earlier at this rate over a period of centuries. References to puberty, to be found in writings of Shakespeare's day, suggest that girls in urban environments experienced menarche at about 14 years of age, and those in rural areas somewhat later. Tanner concluded that, in all probability, 'In the towns of Europe the rate of maturation was slowed down during the late 18th and early 19th centuries; in the villages of the less agriculturally rich countries the rate (of maturation) has always been substantially slower than in England. From the Shakespearian figure of 14 we arrive, about 1820, at the menarcheal age of $14\frac{1}{2}$ for the educated upper classes and $15\frac{1}{2}$ for the less well-off townspeople. By 1960 the figure had decreased to 13 for both groups.'

It has been found that European girls experience their first menstruation in the month of their birthday more frequently than in any other month. This was found to be true when month of birth was compared with month of menarche in 10,117 girls; the association was attributed to 'psychological influences'.[369] It was suggested that a heightened expectancy of the event, operating through the hormonal mechanism, advanced or retarded the changes leading to menstruation. A similar effect has been attributed to other events of significance to the girl, such as a big religious or national festival.

The 'average age at puberty' is perhaps best calculated by the use of 'probits' since each individual can be classified as either having experienced menarche or not, and the percentage of each class can be calculated for successive age-groups. The figure arrived at is therefore the median age at menarche—the age by which half the members of the population under study have experienced it. The method was used in a similar way to determine the median body weight of wild rats at first ovulation.[220] Body weight was used to indicate age because the latter could not be estimated directly. Autopsy revealed whether or not an individual rat had ever ovulated and so each rat could be assigned to one of two alternative classes.

In this and similar work on species with an oestrous, as opposed to a menstrual, cycle the onset of puberty in the female may be taken as the time of the first oestrus. It is more difficult to define or to observe in the male, but in all mammals the changes associated with the attainment of sexual maturity follow the first production of significant amounts of gonadal hormones—oestrogen in the female and androgen (testosterone) in the male.

Changes associated with puberty. The primates show a 'growth spurt' (in height) during puberty. In both sexes the rate of growth increases quite suddenly before it declines to zero as adult height is reached. The 'spurt' lasts about two years in boys and girls; it begins, typically, in the 11th year in girls and nearly two years later in boys. It is more marked in boys, who grow by an average of about 20 cm during the period of accelerated growth. The corresponding average increase in girls is about 15 cm but as girls start their 'spurt' earlier, the 'average girl' overtakes the 'average boy' in height for a time during their 13th year.

The cause of this growth spurt is not known; the 'extra' growth of boys may be due to testosterone but oestrogen is apparently not involved in the spurt in girls. This suggests that most of the acceleration in growth in both sexes is due to some factor other than the sex hormones.

Other bodily changes taking place at this period are directly associated with the production and circulation of oestrogen in the female and testosterone in the male. Examples of such changes include the deposition of subcutaneous fat in girls, and the increase in muscle development in boys. It was formerly thought that the closure of the epiphyses was due to hormones, oestrogen taking effect earlier than testosterone either because it is produced at an earlier age or because it is more effective. This explanation, however, is probably inadequate; it has been suggested that synergism between the sex hormones and growth hormone is involved, in some way not yet elucidated.

Puberty in other species

Several authors have described the occurrence of 'waves' of development and atresia (degeneration) of follicles before the first ovulation in a number of species. In the mouse, for example, some follicles begin to grow about six days after birth.[37] Many of them have traces of an antrum by the 14th day, and they continue to grow until, about 3 weeks after birth, the ovary is as large as that of the adult. During the 4th week, however, most of these follicles degenerate. The largest among those that survive this period continue to grow, and they eventually ovulate about 8 weeks after birth. In the guinea-pig Bookhout distinguished two such 'waves' of follicular development before birth, and three more before puberty, the last of them providing the follicles that rupture at the first ovulation.[33]

No 'growth spurt' like that of man and the higher primates has been recorded in non-primate species, but sexual dimorphism is, of course, often strongly marked, and is controlled by the sex hormones. In male rabbits 'the onset of puberty coincides with the time when the testes become androgenically active, the accessory glands begin to secrete fructose and citric acid . . . and the animal assumes the characteristically male behaviour'.[358] Apparently spermatogenesis was preceded by the onset of androgenic activity—or a great increase in it—and the appearance of sperm in the semen evidently marked the completion of a complex sequence of events. It seems to be generally true that the male gonad secretes a certain amount of androgen from birth or soon after. In the bull, puberty coincides with a qualitative change in testicular androgen production, the output of testosterone being increased at the expense of that of its precursor, androstenedione. Puberty cannot be attributed to a similar metabolic change in other species, however; in the ram, for example, the production of testosterone begins earlier in life, and puberty is evidently associated with an increase in the amount produced.

Female mice reach puberty sooner if they are caged with a male than if they are kept in an all-female group. Solitary animals mature even later.

The effect of the male's presence is most marked during the post-weaning period, but it has some effect even when the males are present up to the time of weaning and are then removed. The effect was not due to more rapid growth, but to the attainment of puberty (first oestrus) at an earlier age and body weight.[376] It is perhaps more natural to regard the absence of males as a retarding influence, rather than thinking of their presence as 'advancing' puberty, but the phenomenon is strongly reminiscent of the 'Lee-Boot' effect, and other similar phenomena (see p. 176), in that an apparently 'social' factor is shown to affect gonadal function and, indeed, to induce oestrus.

Puberty in wild mammals. Many small and relatively short-lived rodents apparently breed, as a general rule, in the spring of the year following their birth, but some, born early in the year, breed before that year is out. The onset of puberty in such animals is, obviously, closely related to the breeding season. A recent study of the Skomer vole showed that although the males apparently never reach sexual maturity in the year of their birth, a few females apparently do so, presumably mating with older males.[81] Puberty corresponds to the onset of the breeding season in the majority of voles on the island, as in mainland populations, and its timing is fairly constant. There is some evidence that the breeding season may be prolonged in European vole populations if there is abundant food in the autumn, but the onset of breeding is evidently less variable. It is perhaps governed by changes in daylength, but there appears to be no evidence on the point. In North America, on the other hand, a considerable difference in body-weight (and therefore, presumably, in age) at puberty has been found in the Californian vole (*Microtus californicus*) in different years, corresponding to differences in rainfall and consequently in nutrition.[149]

Nutritional factors can cause variations in the age at which an animal attains puberty; this is evident both in the domesticated species and in the wild. In studies of wild animals, Sadleir[329] distinguishes two types of investigation: 'either reproduction parameters are compared for populations sampled at the same time but from areas whose major environmental difference is in the level of nutrition available (area comparisons), or such parameters are compared in samples from the same population between years when the level of nutrition varied (time comparisons)'.

Both, as Sadleir points out, have their drawbacks, and both are subject to the limitations usually inherent in studies of wild populations,

F

particularly the impossibility of excluding the influence of factors other than the one it is desired to study. A good example of the 'time comparison' is Newsome's study of the red kangaroo in central Australia.[274, 275] The region is subject to prolonged droughts, and consequently poor nutrition, and these conditions have a very significant effect on the age at which the females reach maturity—about 35 months compared with approximately 27 months in a period of high nutrition after rains. An 'area comparison' is available for elk in Newfoundland, where Pimlott found that only 29% became pregnant as yearlings in an area of relatively poor nutrition, compared with 67% in other areas.[300] A great increase of breeding among yearling deer in Washington state, when the population was reduced by excessive hunting (a 'time comparison'), was also attributed to the resultant increase in available food.[56] Nutritional factors, again, were held to account for the fact that, in New Zealand, red deer may produce a calf by the time they are two years old;[85] in Britain the first calf is usually born when the mother is four years old.

In studying wild animals, it is rarely possible to determine both body weight and age, but mammals in general tend to attain puberty at a body weight that is characteristic of the species. Reduced nutrition delays puberty because it slows growth, but poor nutrition appears to affect age at first oestrus even more markedly than it affects growth, so that puberty occurs at a higher body weight in the more slowly growing animals. This was found to be true of penned wild rabbits studied in Australia[267] and of farm animals throughout the world.[211]

Puberty in female farm animals. All the available information covering the effect of nutrition on age or body weight at puberty in wild animals is based on comparisons, time-based or area-based, between low and high planes of nutrition—that is, on levels of total intake. Work on laboratory and farm animals, however, shows that the effect of nutrition, in this respect, depends very largely on the protein intake. An insufficient supply of protein can delay puberty very significantly even when the total amount of food eaten is normal.

Among farm animals, the average age at puberty differs between various breeds. Most of the available data, however, have been gathered from animals kept under a variety of conditions, and some of the tables that purport to show breed differences do not in fact do so, because the range of variation within each breed is wide and the standard errors of the means are large. However, these surveys and experiments serve, for the most part, to confirm the experience of breeders, which may be relied on, and show that, in general, beef breeds are later-maturing than dairy breeds.

The effects of plane of nutrition on the attainment of puberty within breeds is more clearly documented.[266] In a typical experiment three groups of calves were compared, one fed on a 'normal' or 'standard' ration, one on a ration about 65% of the normal and the third on a ration about 140% of the normal. The average age at first oestrus was 11·3 months in the normal group, 17·3 months in the under-fed group, and 9·4 months in the over-fed group. Comparable results have been obtained with sheep and pigs.

Delayed puberty, or failure of oestrus, has been found to be associated with protein-deficiency (as distinct from low total intake) in cattle as, for instance, in winter-housed animals fed on beet, sugar-beet pulp and straw. A deficiency of phosphorus, or a high Ca/P ratio, in the diet, may also have this effect. In this connection it may be as well to remark that the effect of nutrition on reproduction is not confined to the time of puberty. Indeed, there are few reproductive processes, either during the oestrous cycle or during pregnancy, that are not affected by gross, and sometimes by more subtle, modifications of the animal's diet. Recent work on pigs has shown that the ovulation rate and the incidence of embryonic mortality may be markedly affected by a short period of fasting, or the intake of a single heavy feed at a critical stage in the cycle.[168] This observation, and others of a like kind, shows that the relation between nutrition and reproduction may be extremely subtle. The results of long-term experiments may be confused by the operation of such short-term effects within the operational conditions of the experiment: one is led to wonder whether this sort of effect accounts for some of the contradictions in reports that have emerged from long-term experiments on the effects of high- and low-plane nutrition. The more general aspects of the relation between nutrition and reproduction are discussed elsewhere (p. 169).

The endocrinology of puberty

The ovary, in contrast to the testis, secretes little hormone until the first crop of follicles begins to ripen. Experiments have shown that in the rat sexual differentiation of hypothalamic function is established in the male, under the influence of gonadal hormones, within a few days of birth (see p. 60), but it is not clear whether the ovary has any similar function in very early life. The question therefore arises, whether the initiation of ovarian function (puberty) depends on the availability of gonadotrophin or on the ovary's capacity to respond to it. The ovary's refractoriness to gonadotrophin disappears some time before the first ovulation, and the direct cause of sexual maturation is a rise in the output of pituitary hormones, but the cause of this rise remains obscure. So, when I was asked, during a recent discussion, 'Do you think the periodicity of the cycle is controlled by the ovary or the hypothalamus?' the only concise answer seemed to be 'Yes'.

Donovan has remarked that 'Gonadal activity and pituitary function are in a delicate state of equilibrium during infancy'[106] and a number of experiments have shown that the oestrogen produced by the ovary before puberty, although secreted in relatively minute amounts, nevertheless plays a part in normal development and affects the cell structure of the pituitary gland.

The 'equilibrium' between gonadal and pituitary activity referred to by Donovan, can equally well be considered as a balance between the ovary and the hypothalamus, since the latter controls the pituitary. Puberty is possibly due to a slow change in the threshold of sensitivity of a hypothalamic 'centre' to gonadal hormones. A mechanical model of this concept might be provided by a balanced see-saw, initially held level by equal pressure at both ends and eventually set swinging by a temporary lowering of pressure on one side. They suggested that 'this hypothesis would seem to afford an explanation of the information available at present and does not exclude other factors, such as changes in the sensitivity of the target organs to their trophic hormones, from playing some role in the timing of puberty.'[107]

The first occurrence of ovulation indicates that the secretion of LH has been increased above the threshold level of ovarian response. There is some clinical, but as yet no experimental, evidence that ovulation depends on an increase in LH secretion while FSH secretion remains relatively constant. The clinical evidence is based on the urinary excretion of gonadotrophins, which occurs only in primates. Measurements of the gonadotrophin content of the pituitary gland have provided conflicting evidence, presumably because the amount or concentration of hormone in the gland is not necessarily an indication of the rate of its secretion. The most recent evidence comes from the application of radio-immunoassay methods that are specific for LH and FSH; both hormones are released in increased amounts shortly before ovulation in the rat.

The question whether it is the pituitary (more accurately, the hypothalamus) or the ovary that starts the swing of the pituitary-ovarian balance obviously reminds one of the 'chicken or the egg' puzzle. It would be neatly solved if some other organ were found to intervene at the appropriate time, and so set the escapement mechanism in motion. Such a solution was offered by the suggestion, made some years ago on the basis of experiments with rats, that puberty follows a change in liver function. Gonadal hormones are inactivated before excretion, and there was evidence that the infantile rat liver was unable to inactivate endogenous oestrogen. Donovan and O'Keefe[108] found 'a marked correlation between the change in liver function and the timing of puberty', but they did not consider that the one adequately explained the other, because they found it difficult to imagine that the liver could be directly affected by experimental stimuli such as those which have been shown to be capable of advancing the onset of puberty. They thought it more probable that the stimulus to gonadal activity comes from within the 'hypothalamo-pituitary-ovarian complex' itself. The critical change may be an alteration in the kind of oestrogen secreted by the ovary. The experiments on liver function showed that the infantile rat liver was capable of inactivating pure oestradiol, administered by injection. As it was not capable of inactivating the endogenous oestrogen, this observation suggests that the endogenous oestrogen was not oestradiol. As a working hypothesis the following might be envisaged: oestrogen is regarded as inhibiting the pituitary output of gondaotrophin, and the pubertal increase in gonadotrophin output is regarded as a release from inhibition. The first of these conditions is well established, and the second is quite probable. Thus, during the time that the gonad secretes a type of oestrogen that the liver cannot eliminate from the circulation, the circulating level of this oestrogen will remain relatively high, even though the rate of production is low. When the gonadal oestrogen is changed to oestradiol, which the liver can metabolize, it will very rapidly be eliminated from circulation, the circulating level will fall and gonadotrophin production will no longer be suppressed. From this time on, the predominating ovarian oestrogen will always be oestradiol, but periodically it will be produced in such large quantities, by maturing follicles, that it will affect the hypothalamus in spite of the liver. A change in the kind of oestrogen secreted, from one that the liver cannot eliminate to one that it can, is therefore equivalent to a change in liver function.

Yet another entirely hypothetical possibility is that the critical change is mediated through the adrenal glands, which may be responsible for changes in steroid conformation. For example, steroids are readily sulphated and are physiologically much less active in this form. The adrenals are under hypothalamic-pituitary control in much the same way as the ovaries, so that a hypothesis involving an adrenal mechanism has the attraction of its being presumably susceptible to the influence of external stimuli.

Whatever the immediate cause of the onset of puberty, it seems probable that the ultimate cause lies in the changes involved in the development of the hypothalamus. Such a change in hypothalamic function in the course of development is envisaged in the hypothesis that the hypothalamus possesses both an excitatory and an inhibitory mechanism controlling gonadotrophin secretion, and that the pituitary inhibition, characteristic of the pre-pubertal animal, is due to the lower threshold of the inhibitory as compared with the excitatory mechanism.[107] The small amounts of oestrogen secreted by the pre-pubertal ovary are apparently insufficient to activate the excitatory mechanism but are adequate to maintain inhibition. A change in the relative sensitivity of the two hypothalamic centres would substitute an excitatory effect for the inhibitory one, and so lead to follicular growth and the production of very much larger amounts of oestrogen, which would again have an inhibitory effect.

Damage to the hypothalamus often leads to precocious puberty whether the damage is caused by experimental procedures in laboratory animals or by tumours or accident in children. Several authors have postulated a gonadotrophin-inhibitory centre located in the posterior hypothalamus in man. Its function is envisaged as holding back the secretion of gonadotrophin in early life; thus damage to it would release the inhibition. There is support for this concept in the observation that the intraventricular injection of oxytocin, which stimulates gonadotrophin release, advances the onset of puberty in rats. Precocious puberty has also been induced by oestrogen, administered by systemic injection or in hypothalamic implants, and this result has been attributed to the stimulation of LH secretion by oestrogen. The evidence is inconclusive, for anti-oestrogens have also been observed to cause precocious puberty. In fact, it would seem that if the 'delicate state of equilibrium' in the pre-pubertal hormonal system is disturbed in any way, puberty is more likely to be advanced than postponed. This in itself perhaps supports the idea of an inhibitory centre holding back the release of gonadotrophin until a certain stage of maturity is attained.

Puberty can, however, be delayed by treatment with certain drugs during neonatal life. Italian workers have suggested that puberty is. delayed and a state of permanent di-oestrous induced, in the rat, by treatments favouring the dominance of LH, whereas treatments that favour the dominance of FSH tend to lead to precocious puberty and permanent oestrus. They have accumulated data which support the contention that discrete areas within the hypothalamus are specifically involved in the synthesis of particular releasing factors.[248] The assumption is that certain neurons secrete a particular releasing factor. Neurons secreting FSH-RF are concentrated in the suprachiasmatic area and in the arcuate-ventromedial area. Many of the neurons secreting LH-RF also occur in the latter region, few occur in the suprachiasmatic area, and they are rather more concentrated in the paraventricular area, which does not synthesize FSH-RF.

The localization of specific synthetic activities within the hypothalamus has been referred to in the section on hypothalamic pituitary relations (p. 65).

The pineal body

The functions of the pineal body, or epiphysis, are imperfectly understood, but it is clear that it exercises some degree of control, of an inhibitory nature, over gonadal activity. Puberty tends to be advanced in girls whose pineal body is destroyed by a tumour, or after pinealectomy. A pathological increase of pineal tissue (parenchymatous pinealoma), on the other hand, depresses gonadal function.

The pineal body is part of the brain; in some lower vertebrates it is related (in the way that the optic lobe is related to the eye) to a light-sensitive organ, the pineal eye, situated in a foramen on the median junction of the parietal bones. Such an organ persists in one living reptile, *Sphenodon*, and the foramen is so large in some fossil reptiles that it is thought probable that it housed a well developed third eye. In the mammals, however, there is no trace of this organ and the pineal body must depend on the paired optical retinae for any light stimulation it receives. It is thought that nervous impulses engendered in the retina are conveyed to the pineal body by sympathetic nerve fibres. The response consists of a change in the pineal secretion of melatonin, a blood-borne agent which affects the gonads, probably indirectly. The pineal body is the sole, or almost the sole, source of melatonin; it is said to be the only tissue containing the enzyme that regulates the production of melatonin from serotonin. The 'activity' of this enzyme varies with the oestrous cycle in the adult rat in a way that supports the supposition that the pineal has an inhibitory rather than an excitatory effect. The pineal body has been shown to be necessary to the normal development of the ovary and uterus; changes in the histological constitution and enzymatic function of ovarian, tubal and uterine tissues have been shown to follow pinealectomy in the adult rat. The changes are such as to suggest that the operation causes an increase in FSH.[25]

Pinealectomy (or 'epiphysectomy') has been found to advance puberty in the rat; the administration of pineal extracts delayed it, and injections of melatonin did the same. It has been found that daily systemic administration of melatonin to 22-day-old rats retards puberty, slows pituitary, ovarian and uterine growth, and reduces the secretion of LH. Martini and his colleagues carried out experiments designed to show whether the effect on LH secretion, from which the other effects probably derive, is exerted through the brain or otherwise.[248] They found that melatonin crystals had this effect when implanted in the median eminence of the hypothalamus, but not when implanted in the cerebral cortex or in the pituitary gland. They suggested that it probably acts as an inhibitory signal on specific receptors in the median eminence.

The present state of information about the pineal body has been summarized in general terms as follows: 'The pineal of mammals is not a gland but a "neuroendocrine transducer". It acts to translate a nervous message into an endocrine output. It responds to a sympathetic nervous input by making a unique enzyme which catalyses the synthesis of a pineal hormone, melatonin. Melatonin and other pineal substances act to suppress gonadal growth and cyclicity.[395]

There seems no doubt that melatonin is the active principle of the pineal body, or that it has an 'anti-gonadotrophic' effect, exerted within the brain.[396] A gonadotrophin-inhibiting substance is found in human urine, and was first isolated from that of young males. It was thought possible that this substance might be melatonin or a related compound, but it appears not to be so because it was found that melatonin did not inhibit the ovulatory effect of PMSG or HCG under conditions in which the urinary product did so.[279]

The pineal body contains a relatively high concentration of serotonin which serves as a precursor for melatonin in addition to its other functions. It has been shown that the concentration of both these substances follows a diurnal rhythm, and it appears that whereas melatonin concentration (and production and release) varies in response to exogenous light (photic) stimuli, that of serotonin is 'endogenous' and persists in a 'free-running' state when the photic stimuli are withdrawn or caused not to fluctuate.[396]

The breeding season, and the effect of external factors on ovarian activity

In seasonally breeding species the beginning of each breeding season is somewhat like a repetition of puberty, but it may be envisaged either as a re-activation of an established mechanism, or as an increase in the intensity of a process which has been working at sub-threshold levels for a time. The basic mechanism does not require to be reconstituted each time, and the onset of the breeding season is clearly associated, in many animals, with one or more recognizable factors in the environment. The most important of these is light, usually in terms of day-length rather than daylight intensity. It is certain that its effect on the reproductive system is exerted through the hypothalamus, but how it is transmitted to the hypothalamus, or recognized by this region of the brain, is not yet known. Experimental evidence supports the view that the effect of light is exerted, as one might expect, by way of the eyes and the optic nerve. Direct nervous pathways between the optic chiasma and the hypothalamus have been described, but there is, so far, no experimental proof that the impulse is transmitted by this route.

The problem of how changing daylength affects gonadotrophin output is clearly bound up with the study of circadian rhythm. Rowan's earlier work on the effect of light on gonadal function (later summarised in a review)[317] was published about the time when Smith and Engle[363] demonstrated the functional dependence of the ovary on the pituitary. Marshall related these observations to the regulation of the breeding season, and discussed the implications of the control of gonadal function by 'exteroceptive factors' in relation to the ecology of various species and their adaptation to climatic factors. In his Croonian lecture to the

Royal Society in 1936 he discussed 'Sexual periodicity and the causes which determine it',[245] and in 1942 he reviewed the evidence then available.[246] He, and others, concluded that each species is subject to an inherent rhythm which approximately determines the onset of the breeding season and can be modified by external factors 'in such a way that the young are born at a favourable time of year'.[140] The 'coarse and fine adjustment' of this system is reminiscent of the regulation of the inherent 'free-running' circadian rhythm of sleep and activity, by its 'locking on' to some external event (such as sunrise) perceived through the senses. The parallel between the annual and the daily regulatory mechanisms may be very close, and the term 'circannian' (about a year) has recently been introduced to match 'circadian' (about a day) in discussing events that are temporally related to the seasonal time scale in the way that others are related to the diurnal rhythm.

There is some experimental evidence that a basic inherent rhythm exists and that it is governed, as one might expect, by the hypothalamus. Some degree of sexual rhythm persists, for instance, in rats of which the hypothalamus has been effectively separated from the rest of the brain.

It has also been claimed that an intrinsic seasonal rhythm exists in the sheep, the species in which the effects of light (day-length, latitude, etc.) have been more widely studied than in any other mammal. Sadleir, however, argued strongly against this hypothesis, even on the basis of the experimental results quoted in its support. He writes 'There can be no doubt that the normal cycle of seasonal breeding does persist for a short period of time under altered environmental conditions, but the length of this time is so dependent on the timing of the commencement of the experimental regime that this merely suggests a seasonal pattern of refractoriness. Much of the remaining evidence purporting to demonstrate an inherent rhythm is based on small samples of animals with periods of oestrus whose distribution in time does not stand up to statistical analysis. It is therefore suggested that, to date, all the evidence presented can be more validly interpreted as demonstrating that the seasonal nature of breeding in mammals is entirely dependent on responses to external environmental stimuli and there is therefore no need to postulate any inherent annual rhythm in this group'.[330]

At first glance, this may seem surprising, but it should be remembered that the daily and annual rhythms are by no means truly comparable. Circadian rhythm has been demonstrated throughout the animal kingdom, even down to unicellular organisms. Individual cells may possess this 'free-running clock', as though life had always been conditioned in this way—and there has always been 'alternate night and day'. There is no firm evidence that a corresponding annual rhythm has been incorporated into the genetic structure.

It seems clear from the multiplicity of stimuli that must reach the brain via the senses, and from the fact that lesions, or physical stimulation, of the hypothalamus almost always have a positive effect, that the hypothalamus exerts an important 'integrative' function. It must be more than simply a relay station. It receives a multiplicity of 'messages' from the internal and external environment, sifts them, and 'integrates' them with its own deep-seated rhythms, passing on to the pituitary instructions that are designed to be appropriate to the situation as a whole.

Thus the onset, duration and intensity of the breeding season is controlled through the activity of the gonads, governed by internal and external conditions whose effect is mediated through the hypothalamus. The relative stability of their external environment accounts for the fact that many laboratory and domestic animals breed all the year round. Even in these conditions some degree of seasonality remains in many species. Particular factors can be isolated in laboratory conditions, and effects have been observed in this way that might never have been recognized among the complex interactions that operate in wild populations. In this section I propose to describe some of the modifications of ovarian activity imposed by changes in the external environment, beginning with the effects seen in individuals under laboratory conditions and proceeding to the more general responses that make up the breeding behaviour of a whole community or population.

In a close study of seasonal ovarian changes in the sheep, the polyoestrous domestic sheep was compared with wild species that experienced oestrus only once in the breeding season.[219] The ovaries of sheep killed during each calendar month of one year were weighed, and a significant fall in ovarian weight was found during March and April, and a less marked fall in July and August. Similarly, the ovaries reached maximum size in November and December, with a lesser increase in May and June. The weight gain was accounted for, in each case, by an increased number of developing follicles. The decline after each peak was caused by an increase in the number undergoing atresia. Thus there are two periods in the year when the ovaries contain a peak number of follicles near maturity. The number of corpora lutea was highest in the period from November to January. The minor period of ovarian activity perhaps suggests the evolution of a restricted breeding season from a more extended one.

The relationship between the breeding season of domestic species, and that of their wild progenitors, offers scope for speculation. Some may have evolved towards a more extended season, others towards a more restricted one. As more information becomes available, the concept of rhythmic variations in the intensity of a continual activity becomes increasingly firmly established, in preference to an 'on-off' pattern of alternating activity and inactivity.

Ecological factors affecting breeding

Sadleir's recent book on the ecology of reproduction[330] provides an extensive coverage (and about 600 references) of this field. Ovarian function is to some extent involved in every aspect of reproductive activity, and ovarian physiology is affected by and in turn affects every ecological phenomenon. It would, however, be unprofitable to take ecological examples one by one and attempt to explain them in terms of known physiological modalities. It could be done at a hypothetical level, but equally it could have been done long ago, when the hypotheses were simpler. The primary value of Sadleir's book lies in the orderly way in which the wide range of ecological observation has been brought together. In his treatise, the breeding season naturally figures large, and the ecological factors affecting it are considered under the headings: light, temperature, rainfall and nutrition.

Photoperiod. The importance of light, particularly day-length, in determining the onset of the breeding season, has already been emphasized. Among the farm animals the duration, and timing, of the breeding season is of particular importance in sheep. Generally speaking, the breeding season, at least among British breeds of sheep, centres around the shortest day (winter solstice). It is longer in some breeds than others, the difference lying mainly in the time of onset. Hafez showed that the duration of the breeding season was related to the latitude and altitude of the place of origin of the breed, being shorter in breeds from regions nearer the poles.[153] He was not able to relate the onset of oestrus to any specific characteristic of the light/dark curve, much less to suggest a physiological pathway by which the environmental change affects the reproductive organs. He did show, however, that when sheep were kept in continuous light, or under 'long-day' conditions, from the onset of the breeding season, breeding ceased many weeks earlier than the normal time.

It is equally certain that the breeding season of the mare is affected by the photoperiod, but again it is not clear whether the effective agent is the amount of light, the length of darkness, the light dark ratio, or the cumulative rate of change. Individual mares vary greatly, but in general the breeding season starts in early spring (February to April) and ends in autumn (October to December).

The influence of light on the reproductive system has been fairly extensively studied in several species of laboratory animal since Bissonnette, in 1932, showed that additional illumination would induce oestrus in anoestrous ferrets. Bissonnette also attempted the converse experiment by placing ferrets, in the breeding season, under winter daylength conditions. The results were inconclusive, and later work has shown that the ferret appears to have an inherent rhythm of sexual activity, and that this rhythm is modified by changes in the light/darkness ratio of the diurnal cycle. It has been shown that 'long-day' conditions of lighting may advance the onset of breeding (oestrus) in ferrets as compared with natural day-length or a constant 9-hr day.[163] A 6-hr day retarded it, but in both ferret and mink the gonadal response to daylength is far from simple. The majority of the animals used in these experiments apparently showed an 'inherent rhythm', but the possibility that it was due to seasonal temperature changes could not be entirely excluded.[162] Daylength changes also affect the deposition of body fat and the replacement of fur in the ferret, and in the mink. In the latter, and in the pine marten, the duration of pregnancy may be altered by daylength changes; this is due to control of the time of implantation after a period of 'delay'. The administration of progesterone does not, by itself, have this effect, and Hammond has postulated the existence of an extra-ovarian factor acting on the uterus and regulated by light.[161]

In contrast to its effect on the ferret, exposure to continuous light eventually induces a persistent oestrous condition in the rat. This has been known for more than 30 years, and such animals have been used fairly extensively in studies of the action of hormones in inducing ovulation. The mechanism of the effect remains obscure. It is evidently linked with the establishment of a circadian rhythm in ovarian function, and it seems that this rhythm serves as a 'fine control' which governs an inherently rhythmic activity. That the rhythmicity of ovarian function is inherent, and that the circadian regulation is superimposed, is indicated by the fact that if rats are put into conditions of continuous darkness at an early age (soon after the oestrous cycle has been established) they continue to have regular cycles. If, on the other hand, the cycle is given some weeks or months in which to become 'locked on to' the circadian rhythm before the rats are placed in the 'dark' conditions, the oestrous intervals are soon prolonged and the cycle may cease altogether.[197]

The effects of light and temperature, and their interaction on gonadal activity, have also been studied in the field vole. In three experiments using a laboratory-bred stock, temperature had little or no effect, either independently or by interaction with light. In a fourth experiment, using animals caught in the wild, both temperature alone and the interaction of light and temperature exerted a significant effect.[71] The authors speculated cautiously about the possible significance of the difference, and suggested that wild conditions would eradicate, by natural selection, a genetic make-up that did not respond in this way, whereas a laboratory stock could tolerate it. This particular account illustrates the complexities of ecologically significant interactions between various environmental factors. The usual experimental method of studying the effects of a single factor is by isolating it; relatively little progress has yet been made in studying the multiplicity of interactions that determine events within individuals and so control the breeding season and, eventually the population cycle, in whole communities.

Temperature. Before it was realized that light changes, particularly changes in daylength, were of primary importance in controlling the reproductive behaviour of mammals and birds, temperature was believed to be the principal controlling factor. Temperature clearly does affect breeding, but it has been shown to play a secondary role in most cases. It does not necessarily affect ovarian function directly. An interesting case in point is that of rats (*R. norvegicus*, the brown, Norway or sewer rat) living on garbage dumps near Nome, in Alaska.[334] This is the most widely distributed species of mammal, man alone excepted, and it survives and thrives in an astonishing variety of habitats. I have observed a colony that lived and bred in the roof-space of a building in Nahalal, near Haifa; the rats were active during the hottest part of the day, when the temperature in the roof space reached $45°C$.[385] The brown rat normally breeds throughout the year but the rats at Nome, at the extreme northern limit of the *Rattus* range, were found to cease breeding for a few weeks at the coldest time of year. The evidence suggests, however, that this was not due to a prolonged anoestrus in the females, but to the cessation of spermatogenesis in the males, the testes being withdrawn into the abdominal cavity.

The ubiquity of the brown rat, and its survival in extreme climatic conditions, is evidently related to its adaptability to a variety of types of habitat. Because it can use almost any kind of cover it can spend much of its time in a microclimate that must be more comfortable than the external conditions would suggest. The rats at Nome were frequently found to suffer from frostbite, but it is unlikely that they spent all, or even much, of their time in temperatures below freezing. Garbage generates heat, and the rats were probably able to burrow in warm material, experiencing extreme cold only on foraging expeditions. Nevertheless, they did get frostbitten and their pregnancy rate dropped almost to zero. In contrast to these frostbitten rats, those living on garbage dumps in Britain have been found to suffer severe burns, presumably due to their being overtaken in their burrows by the slow combustion that frequently occurs near the tipping face.

Rats and mice, particularly the brown rat, *R. norvegicus*, and the house-mouse, *Mus musculus*, are remarkable in their close association with man, and the consequent variety of 'unnatural' situations in which obviously 'wild' populations exist for long periods. Perhaps the most astonishing example is that of the cold-store mice first described in the course of work on rodent infestation during the second World War. The cold-store temperatures never rose above $-10°C$; the mice lived permanently in them and were characterized by heavier body-weight and larger litters than mice in other urban habitats. The population as a whole was no more 'productive', apparently because their larger average litter size was offset by their not beginning to breed until they had attained a greater body weight than mice in other habitats.[217]

These cold-store mice survived by constructing their own 'microhabitat'. There were no warm corners in the stores, but the mice built nests of insulating material from the hessian bags and fowl feathers that were available from time to time. Thus, although they had no external source of warmth, they achieved a degree of insulation, for much of the time, that was well above the insulation provided by their own pelage.

These observations were followed up, and greatly extended, in a long series of experiments on the breeding of mice at different temperatures.[26] The observations are interesting and sometimes surprising. Growth was slowed in cold conditions, and puberty was delayed because of this, as in other situations where puberty appears to depend on body weight rather than age (see p. 153). The unmated cycle (i.e., the dioestrous interval) was longer in mice reared and kept at $-3°C$ than in controls, but mice reared at $21°C$ and transferred to cold conditions as adults recovered their normal cyclic ovarian activity after an initial period of prolonged cycles. In all these experiments the 'cold' mice were provided with insulating material, but their reproductive performance was, in most respects, impaired. There were differences between inbred strains, and it was found that the advantage of hybrid vigour (heterozygosis) was apparently enhanced in the adverse environment. For example, the F_1 hybrids of C57BL and A2G strains reared twice the weight of young produced by either pure strain when kept at $21°C$, but at $-3°C$ they produced nearly five times the weight produced by the parent strains.

There are many well documented accounts of the breeding season of wild mammals being curtailed or extended, at either end, by high or low temperatures, but the manner in which this is brought about has not usually been determined.

A special case of the effect of ambient temperature on reproductive activity is provided by the true hibernators. These, as distinct from such animals as the badger, which tends to 'sleep-in' in winter, are species in which metabolic changes occur in hibernation. They are probably restricted to certain genera of bats, insectivores and rodents. In them, the basal metabolic rate (BMR), and the deep body temperature, fall below the normal level during hibernating sleep. The metabolic changes are such as would be expected to result from the degree of hypothermia experienced. In effect, the animals appear to 'switch off' their homoio-thermic mechanism and become temporarily 'cold-blooded'. None of them remain in this state continuously for really long periods—probably never for more than about ten days. At intervals they awaken, the body temperature and BMR rise, and the animals move about and feed.

Some of the true hibernators have a breeding season similar to that of related species that do not hibernate, that is, they enter anoestrus before hibernating, and remain anoestrous for some time after the spring awakening. This is true of some rodents, and perhaps of the hibernating shrews. The peculiar cycle of some bats, in which ovarian activity is arrested in hibernation, in a sort of prolonged pro-oestrus, is referred to elsewhere (p. 87). There is some evidence that, in hibernating rodents also, the reproductive processes 'prepare beforehand' for the end of hibernation. It is clear that ground squirrels (*Citellus*) mate very soon after they finally awaken, whether this is early in March, as observed in some years, or not until April. But spermatogenesis and follicular development have both begun before the animals enter hibernation. As in the case of the bats quoted above, the result is the early appearance of young soon after the spring awakening.

Nutrition. It is a truism to say that, in the wild state, the search for adequate food and water dominates an animal's life. In the mammals, including man, purposive behaviour is largely governed by nutritional requirements, under the driving force of hunger. The sensation of hunger is induced in man, and apparently in mammals generally, by lowered blood sugar, and this appears to be the cause of hunger in natural conditions. As, in most species, the normal level of blood sugar can only be maintained by frequent replenishment of the carbohydrate stores, hunger is a constantly recurring, if not actually continual, 'motive for action'. Failure to alleviate this sensation is itself a 'stress', which affects the endocrine system. Continuing failure of course leads to inanition and death, but in the intervening stages, when nutrition is sub-optimal, some essential bodily functions are, as it were, given priority. During pregnancy, the foetus is so favoured, at least in later stages of gestation. The effects of nutritional shortage, as distinct from normal hunger, conceivably operate through the hypothalamo-pituitary complex, or on the metabolism of sex hormones, or on the responsiveness of their target organs. The bulk of evidence is that the first is by far the most important route, and 'where effects are claimed on glands other than the pituitary they can often be explained by the effects that lack of nutrient has had on the pituitary itself'.[214]

McClure[235] showed that a brief period of fasting, imposed on mice about the time of mating, may lead to death of the embryos before implantation. Since the effect was prevented by the administration of progesterone, or gonadotrophins, he concluded that it was due to ovarian hormone failure resulting from pituitary gonadogrophin failure. Later work has shown that the effect is probably due to suppression of the hypothalamic releasing factors. The gonadotrophin content of the pituitary gland may actually be increased, presumably because its synthesis continues beyond the time when its release is inhibited.

The effects of protein deficiency are very similar to those of an overall deficiency, but lack of various specific mineral constituents has been shown to affect the reproductive cycle, and vitamin A deficiency causes degenerative changes in the female reproductive tract. These changes include metaplasia (overgrowth) of the uterine epithelium, and it has been found in rats that this metaplasia depends on the presence of oestrogen; it does not develop in ovariectomized animals.

Dietary factors of diverse kinds, quantitative and qualitative, so interact with each other and with a wide variety of other factors that it is difficult to isolate particular causes and effects, and it is impossible to compare the results of work carried out in widely differing conditions, even in the same species. Nutritional requirements are so fundamental that any deficiency is in some degree an abnormality, and dietary 'effects' are almost entirely harmful effects.

Sadleir has emphasized the difficulty of evaluating the role of nutrition in the reproduction of wild mammals, due to the problems involved in assessing variations in quantity and quality of available food in natural conditions. He was, nevertheless, able to review literature concerning the role of nutrition in controlling reproductive activity in a wide range of species from a variety of habitats.[329] Concerning the breeding season he remarks that 'the onset and cessation of breeding is markedly affected by the level of nutrition, especially in those environments such as tropical and arid to semi-arid areas where this level fluctuates considerably. In the majority of situations, it is controlled by the incidence of rainfall which governs the growth of new vegetation'. The flush of new growth, after the break of a drought, is the main proximate factor in stimulating breeding, usually after a lapse of time during which the animal's physical condition improves. Exceptionally, the response may be more immediate, as in the rabbit (*Oryctolagus cuniculus*) in Australia, in which oestrous activity has been observed within a week after rainfall following a period of drought. The rainfall itself appeared not to be the direct cause (although such a response has been reported in desert birds) but it is possible that oestrus was induced by oestrogenic compounds in the newly sprouting grass. In most species in desert conditions, it would appear that nutrition acts as an ultimate factor in controlling the time of breeding, while photoperiod acts as the principal proximate factor.

In this context, 'ultimate' factors are the overall factors to which the population is broadly adapted, while 'proximate' factors contribute the immediate cause releasing a particular reaction in the individual. The distinction is somewhat like that between strategy and tactics, where some items may be classed under either head. In anthropomorphic terms, one might read a statement such as: 'Sheep mate near the winter solstice so that lambing will occur when feed is plentiful'. As it stands the statement is hardly permissible, but one may argue that the winter solstice is a proximate, and nutrition an ultimate, factor in determining the time of mating in sheep. Incidentally, one may note that the English language is so dependent on punctuation and word-order that the implication of purposive behaviour in the quoted statement could be increased or removed by changes of this kind.

One can readily imagine that breeding will be encouraged by a sudden improvement in diet, and in the ease with which food is obtained; it is more difficult to envisage the mechanism by which the time of mating is so modified that the young are subsequently born, or weaned, at a favourable season in particular localities. This seems to be achieved in an island population of the quokka, a small Australian marsupial. Mainland populations of the same species breed continuously and the difference has been ascribed to nutritional factors. Many investigations have demonstrated the modification of breeding activity in apparent adaptation to food availability, and 'undesigned experimentation' provides several examples. This refers to comparisons between, for example, a population feeding on natural vegetation and a nearby population feeding on agricultural crops, or on a municipal rubbish dump. Even the grizzly bears of Yellowstone Park haunt the dump and modify their habits in response to a generous supply of garbage left by tourists.[82] Among the island quokkas just referred to, seasonality of breeding was less marked in a group that occupied a rubbish dump than in the rest of the population. Planned comparisons of a comparable sort have been made in work on domestic animals and, like experiments with daylength, they refer mainly to the sheep as far as seasonality is concerned. Different nutritional levels have been found to cause differences in the onset of breeding in sheep. Such differences could be detected nearly a year after the regimes that caused them had been discontinued, although control and experimental animals had meanwhile outgrown the difference in body weight caused by the earlier differences in diet.[362]

Climatic factors. Equatorial animals experience minimal seasonal variation in daylength so that seasonality of breeding, when it occurs in them, presumably depends on other factors. Ecological studies of a wide range of species in equatorial Africa show that it does occur and that it is ultimately determined, in most cases, by rainfall. The ecological implication of adaptability in breeding habits was well expressed by Jarvis:

'It is probable that the ability to adapt the breeding habit to fit the local rainfall pattern is a limiting factor for many small African mammals, and the greater their plasticity in this respect, the larger will be their geographical distribution'.[207] An interesting example of this principle is provided by the hyraxes, which are widespread in Africa and the Levant. They exhibit a 'fundamental plasticity' in their breeding behaviour. 'No single factor has a predominant effect throughout the wide geographical range and it would seem that different environmental factors influence breeding in the tropics from those that operate in the temperate extremities of the group's range'.[331] At the northern and southern extremities, where seasonal fluctuations of temperature and daylength are considerable, temperature and rainfall are the main operative factors, while daylength is a subsidiary factor. In the equatorial part of the range, temperature and daylength are nearly constant, and breeding activity is modified in response to rainfall and the consequent changes in vegetation. In this group (they are divided into several species, but they are evidently closely related) the time of breeding appears to be adjusted so that lactation and weaning coincide with the times of most favourable conditions of diet. Relatively close neighbours, living in the Rift Valley or outside it, live under very different climatic conditions, particularly as regards rainfall, and this is reflected in distinctly different patterns of breeding. It is, of course, impossible to separate climatic from nutritional factors but it is common, and convenient, to regard a population as adapting to particular climatic conditions.

An interesting example of the interaction of climate and nutrition on breeding activity is that of two species of ground squirrel, *Citellus beldingi* and *C. lateralis*. They are found in western North America and their geographical ranges overlap, but they occupy different ecological niches. They are both hibernators, and breed soon after emerging from hibernation, but *C. lateralis* is active for longer and has a longer breeding season. The difference has been attributed to a difference in food habits.[236] The shorter breeding season of *C. beldingi* is to some extent compensated for by a larger litter size, but it is not clear whether this too is related to the diet.

The rabbit's almost immediate response to a climatic stimulus (rain) in Australia has already been mentioned. This species appears particularly adaptable in this respect, and breeding reaches a peak between January and March in New Zealand, but between May and August in the Macquarie Islands and at other times in various parts of the world. The hare, on the other hand, breeds over an extended period in the summer months, mainly from March to June in the northern hemisphere, September to December in the southern hemisphere. When the six-month calendar adjustment is made, the peaks coincide fairly closely over a wide geographical range as, for example, in the Antipodes, Russia, Canada and Scotland.[133] The clue to the difference between the rabbit and the hare in this respect appears to lie in the restricted breeding season of the former and the extended one of the latter. The rabbit evolved in the Mediterranean and has spread, or has been introduced, into many countries. Hares, on the other hand, have an enormous natural range 'from Spain to China, and from Russia to South Africa'. If it were the hare that varied, and the rabbit that appeared conservative, with regard to breeding-season, this argument might seem logically satisfying; as it is the situation merely exemplifies the complexity of physiological and ecological interactions and the inadequacy of over-simple explanations.

Jarvis, in the paper already referred to, describes the breeding habits of a fascinating group of animals, the mole rats, in equatorial Africa. They are very dependent on local rainfall, which affects not only their food supply, but the condition of the soil in which they burrow. One genus, *Heterocephalus*, lives in a relatively harsh climate, with very scanty 'short rains' in November and December and 'long rains' from March to May. Breeding appears to be confined to the period of the long rains and a little time afterwards. During these rains the population is engaged in collecting food and storing it in extensive burrows. Pregnant and lactating females, and juvenile animals, appear to depend entirely on the stored food.

The role of the senses in mating behaviour

Australian biologists have studied the mating behaviour of sheep with particular regard to the means by which a ram can seek out and recognize a ewe in heat, and by which the oestrous ewe can locate a potential mate. This phenomenon of mutual recognition is so widespread, and the mechanism so reliable, that it has largely been taken for granted, but it has a direct bearing on practical husbandry when, as in the case of sheep, animals are kept under range conditions with a very low proportion of males to females. In Australia, it is usual to allot two or three rams to every hundred ewes in a flock. Mating is normally successful even when the flock is spread over a large area, and it is evident that 'contact between ewes and rams is not by chance, but results from the action of specific behavioural patterns'.[132] The senses of sight, hearing and smell were studied by depriving individual animals of one of these senses. Blindfolds were used to eliminate vision when required, but surgical means were used to eliminate auditory or olfactory stimuli, rendering the animal either deaf or anosmic. Sexual activity as such was measured by allowing rams and ewes to mix freely; partner-seeking was measured by tethering the ram or the ewe.

The results suggest that contact between ram and oestrous ewe depends chiefly, but not entirely, on the ram seeking out the ewe. The ram sought out an oestrous ewe mainly by scent, whereas the ewe sought a partner mainly by sight. The various impediments had a measurable effect on the animal's own sexual activity: that of the ram was slightly impaired by deafness, more so by anosmia, and very greatly by being blindfold. The incidence of oestrus was significantly lowered only in the anosmic ewes although, as just stated, the oestrous ewe depended more on vision than scent to locate a mate. This latter observation is difficult to assess, for the 'random' movements of the blindfold animals were restricted, and failure to find a mate may have been due to the blindfold ewe's covering little territory rather than to her being unable to recognize the ram when within normal sighting distance of him.

An interesting experiment bearing on the sensory pathways involved in sexual behaviour and its relation to ovarian function employed the technique of operant conditioning.[255] A male monkey, trained to press a lever to obtain food, was put into a twin-compartment cage equipped with a lever that, when pressed 250 times, activated a servo-motor and raised the dividing partition between the two compartments. At first the 'reinforcement' was food; later, another animal. Two males were chosen as consistent performers, and each was placed in a twin cage the other compartment of which was occupied by one of three ovariec-tomized females. One of these received subcutaneous injections of oestradiol throughout the experiment and both 'trained' males regularly performed the required stint to gain access to her. They did not respond in this way to reach one of the other spayed females, until oestrogen was administered to them. Oestrogen was then applied intravaginally, in sufficient amount to render the female 'attractive' without altering her sexual skin or overt behaviour. When this had been established, the males were rendered anosmic and were found not to respond in this condition. The experiment therefore provides strong evidence that the males were stimulated by 'the olfactory detection of an oestrogen-dependent vaginal pheromone'.

There can be no doubt that olfactory stimuli play a major part in sexual behaviour in mammals, and in many other animals. Special interest attaches to their role in the higher primates, for these, like man, have a menstrual cycle as distinct from an oestrous cycle, and the phenomenon of 'heat', in so far as it exists at all, is very much less obvious in them. The experiment described above was so designed as to allow the macaques' behaviour to be described in quantitative terms. This was also done, but in a very different way, in Birmingham experiments on the same species.[184] Sexual behaviour in this monkey is to a large extent stereotyped in both sexes, but more particularly in the male. By patient observation and scrupulous recording, it was found possible to construct a 'sexual performance index' (SPI) for the males, and to observe the effect on it of hormones administered to the female. The SPI of the male was raised when oestrogen was administered to the female but it was noted that it did not significantly raise the frequency of 'invitation gestures' or 'presentations' made by the female. Its effect, as in the conditioning experiment, was to enhance the female's sexual attractiveness, not to increase her receptivity. This work was carried further when it was found that testosterone, administered to the female, caused a significant increase in the frequency of female 'invitation gestures', but did not affect the male's response. Thus, it would appear that the female macaque's sexual receptivity is affected by testosterone, and not by oestrogen, whereas her sexual attractiveness is enhanced by oestrogen and not by testosterone. Administration of an anti-androgen lowered the frequency of 'presentation' by spayed females, and testosterone restored it. The suggestion has been put forward that oestrogens and androgens function in this way in the normal course of events, and it is postulated that the androgen in the female is from the adrenal glands. This would neatly account for the continual receptivity of the female monkey, together with the increase of mating frequency that has often been reported near the time of ovulation, when a maturing follicle is secreting an increased amount of oestrogen.

'Social' factors in reproductive activity

Under this heading we may include any modification of ovarian activity caused by the presence or the behaviour of other individuals of the same species. In the broadest sense this definition covers all the integrative behavioural responses in herd animals,[154] and the density-dependent effects described in some other populations.[69, 70] A direct effect of the introduction of a ram to a flock of ewes was reported by Schinckel,[335] who showed that when this was done at the beginning of the breeding season, the ewes tended to enter oestrus synchronously. The observations further showed that 'this effect is only seen in the transition from the non-breeding to the breeding season and that the introduction of the ram has no effect on incidence when the breeding season is established as a result of the operation of other factors'.

Pursuing this investigation, Schinckel found that the immediate response of the ewes was, usually, a 'silent heat' (ovulation without oestrus—see p. 86) and he suggested that this accounted for the 'lag' of about 14 days between the ram's arrival and the onset of true oestrus in the ewes.[336]

This work was followed with interest in South Africa and an experiment was designed to provide direct confirmation of Schinckel's observations and conclusions.[201] Possible confusion due to differences in diet was also taken into account in this experiment because other South African work had already shown that the spontaneous occurrence of oestrus at the beginning of the breeding season may be influenced by the level of nutrition at the time. A '2 × 2 factorial' experiment was therefore planned. In this type of experiment, the simplest form of factorial, the interactions between two pairs of factors are compared. There are four groups, so that if one pair of factors is Aa, and the second pair Bb, the quantitative results can be arranged like a game of noughts and crosses (see p. 177).

This experimental plan permits a convenient statistical analysis and when the four groups were compared, it was found that the dietary factor had no significant effect on the occurrence of oestrus. When 'teaser' rams were introduced to the flock on day 3 (day 0 was the date of the ewe's entering the experiment), ovulation occurred on 'day 6·2'. When the teaser ram was introduced on day 14, the average interval from day 0 to ovulation was 18·5 days. As the ewes were randomly assigned to treatment groups, this means that the presence of the ram advanced the date of the first ovulation of the breeding season by an average of 12·3 days, a statistically significant effect.

Among laboratory animals, three well-defined phenomena have been described in which ovarian activity is altered by 'social' factors. These phenomena are commonly referred to by the names of the original observers as the 'Lee-Boot effect', the 'Whitten effect' and the 'Bruce effect'. All three relate to the laboratory mouse and the original descriptions date from around 1955–56.

Van der Lee and Boot, working in Amsterdam, found that spontaneous pseudopregnancy occurred more frequently among female mice housed in groups of four than among mice that were housed singly.[381]

	A	a
B	A B	a B
b	A b	a b

In the experiment under discussion day 0 was the date the experiment began and the factors were 'feeding' and 'teasing':

A—supplementary ration added to diet from day 3;
a—no supplementary ration;
B—vasectomized rams with ewes from day 3;
b—vasectomized rams from day 14.

A 'spontaneous pseudopregnancy' implies the interpolation of an active luteal phase between successive ovulations, without cervical stimulation by mating or by experimental means. Lee and Boot recorded the duration of 939 cycles among a number of mice of inbred strains and their F_1 hybrids. Nearly two-thirds of the cycles were of 4–5 days, very few were of 7 or 8 days, but more than 12% of the total were 10 to 12 days in length. They showed, moreover, that during the lengthened cycles the vagina was mucified and the uterus was capable of a decidual response—that is, deciduomata were produced after traumatization of the uterine horns. They therefore concluded that 'the hormonal situation in the prolonged cycles was comparable with that found in pseudo-pregnancy induced by mating with a sterilized male'. Having observed that 'the manner in which the animals were housed appeared to be an important factor' the authors proceeded to clinch the matter by recording the incidence of pseudopregnancy among female mice of an inbred strain (57BL). No pseudopregnancies occurred in 240 cycles among mice housed singly, whereas 62 (26%) occurred in 242 cycles among mice housed four to a cage. The original account implies that the effect was even more marked when the mice were housed four to a cage after being kept solitary for a period of time. The authors were able to discount the possibility that the pseudopregnancies were caused by the manipulation involved in taking the vaginal smear. They also suggested that 'direct bodily contact between the animals is not indispensible' in causing these pseudopregnancies. The possibility that pseudopregnancy was caused by vaginal or cervical stimulation, by one female mounting another, was therefore ruled out.

Van der Lee and Boot did not discuss, or experiment on, the effect of 'crowding' on the cycle, but they recorded the dimensions of the cages they used, 17 × 11 × 12 cm. Following up their work, Whitten in Canberra found that the di-oestrous interval was prolonged in mice housed in groups of 30 in cages measuring 46 × 31 × 15 cm. This is almost exactly the same floor area per mouse as that provided in Lee and Boot's cages, but Whitten found that the interval between ovulations in his mice was a true di-oestrus and not a pseudopregnancy.[389] He showed that the effect was not due to visual or tactile stimuli. He suggested that the difference between his result (anoestrus) and that of Van der Lee and Boot (pseudopregnancy) might have been due to genetic differences between their respective strains (Whitten's mice were principally of the Walter and Eliza Hall strain). With this caveat, however, he postulated a more constructive explanation on an endocrinological basis. Everett had earlier suggested that FSH inhibits prolactin secretion, and that pseudopregnancy in mice and rats results from a depression of FSH secretion and a resultant abundance of prolactin (luteotrophin: see p. 95). Whitten therefore suggested that 'the effect of grouping observed by Lee and Boot results from an inhibition of FSH secretion. By intensifying the conditions producing this inhibition (thirty instead of four per cage) further depression of gonadotrophin might be expected'. This, he argued, might have produced the anoestrus he observed.

We will return to the problem of the mechanism, or mediation, of these phenomena after briefly describing the 'Whitten effect' and the 'Bruce effect'. The former is quite distinct from Whitten's work on the Lee–Boot phenomenon described above and, like the 'Bruce effect', it involves the male. Briefly, the 'Whitten effect' consists in the tendency to synchrony, induced by the introduction of a male, in the oestrous cycles of female mice that have previously been housed either singly or in groups, but away from a male.[388] The effect is very similar to that described by Schinckel in sheep. The mice he used normally had an oestrous cycle of 4–6 days duration, but after each female mouse was 'paired' with a male significantly more than the expected number of them entered oestrus on the third day, and correspondingly fewer on the first and second or fourth and subsequent days. Oestrus therefore appeared to have been retarded in some and advanced in others. In this paper, in which the phenomenon was first described, Whitten suggested that the effect might be similar to that of an injection of progesterone as described by Everett, using rats, in 1948.[124] In these experiments, the hormone retarded or advanced the time of oestrus, depending on the stage of the cycle during which it was administered. Whitten later showed that the incidence of 'abnormal' cycles (i.e., di-oestrous intervals of more than 4–6 days), among female mice housed in groups of thirty, was greatly reduced when a male was introduced to the group, even if the male was confined within an inner cage. He concluded that the synchronization of oestrus, and the suppression of the long di-oestrous intervals in grouped females, were both caused by olfactory stimuli from the male.

The 'Bruce effect' refers to the demonstration by that author that either pregnancy or pseudopregnancy could be 'blocked' at an early stage by replacing the 'stud' male (the one with which the female had mated) with a 'strange' male (a different one of the same strain) or with an 'alien' male (one of a different strain). Females exposed to strange or alien males within four days of mating with the stud male would experience neither pregnancy nor pseudopregnancy, but would return to oestrus within 3–4 days.[49] As in the case of the Lee–Boot effect and the Whitten effect, actual contact between the sexes was not necessary to the Bruce effect.

The first detailed account of Bruce's observations was published in the first issue of the *Journal of Reproduction and Fertility*. It is clear from this account that the phenomenon was recognized in the course of experiments on the effect of steroids on pre-implantation stages of pregnancy in mice. The account opens with the words: 'Experiments on the effects of feeding certain synthetic compounds to non-pregnant female mice . . . were extended to include the pregnant animal. This involved placing the recently mated female with a different male on the day following the finding of the vaginal plug. Pregnancy failed in a number of mice . . .'. The discovery could not be said to be accidental, because it was made in the course of a planned experiment. The pregnancy block might have been ascribed to the effect of the steroid compounds administered, but it was noted in controls that were receiving the vehicle solvent alone. Sir Alan Parkes, who collaborated with Bruce in subsequent work on this phenomenon, has remarked that 'It is salutary to realise that this most important observation could have been made any time in the last forty years by anyone who possessed some mice and an old microscope.[283] By 'anyone', Parkes of course implied anyone as methodical and thorough as Hilda Bruce. These qualities were amply demonstrated in the experiments in which the original observation was followed up. One of the first of these showed that there was no superfoetation; it was not, therefore, a case of a second lot of ova being fertilized while those of the first litter were retarded. When the stud male was of strain 'A' and the strange male of strain 'B', all the offspring born were of strain 'B'. Next, Bruce investigated the effect of 'proximity without contact, the recently mated female being placed in a small cage inside a stock box containing other mice' so that 'the animals could see, hear and smell, but not touch, one another. Pregnancy was blocked when the stock box contained males; it was not blocked when the mated female was surrounded by other females'. The fate of the 'original' blastocysts was uncertain; it seems likely that they would be expelled from the uterus during the oestrous phase induced by the presence of the strange male.

As pregnancy was as effectively blocked by a strange male when the whole experiment was carried out in darkness, or when the mated female was placed not in company with a strange male but in a box previously occupied by one, it seems very probable that the stimulus involved is mainly, if not entirely, olfactory. Further evidence to this effect was derived from experiments showing that the block did not occur in mice from which the olfactory bulbs had been surgically removed. The immediate cause of the pregnancy block evidently lies in the corpora lutea failing to become functional, and it was found that daily injections of prolactin (known to be luteotrophic in mice) almost entirely prevented the pregnancy block.

An effect exerted on a mammal by something from an external source, and not mediated by visual, tactile or auditory stimuli, must presumably be of a chemical nature, a 'pheromone', detectable by smell or taste, and Bruce has subsequently shown that the effect is abolished by castration. Methods for the recognition and assay of an 'oestrus-accelerating pheromone' in mice have been described; it has been isolated from urine taken from the bladder of intact males or androgenized spayed females, but was not found in urine from castrate males. The urine was free from accessory sex gland secretions, and the pheromone therefore appears to be androgen-dependent, and probably of testicular origin.[47]

Richmond and Conaway have recently reviewed the Lee–Boot, Whitten and Bruce effects.[311] They suggest that the changes in environment and the disturbance associated with the experimental procedure may account, at least in part, for synchrony and for pregnancy block. Thus some experiments with inseminated wild house mice, while confirming the occurrence of the Bruce effect, showed that changing cages, without other factors, could produce as great a degree of synchronization of cycles as cage-changing together with treatment with male urine. In Bruce's experiments it was found that pregnancy block was most effective when the females were transferred to soiled cages twice rather than once daily. Cage-changing does not account for Bruce's results, however, since the block occurred when new cages had been occupied and soiled by a strange male and not when it had been soiled by the stud male. In fact, as Bruce expressed it, 'a high degree of discrimination is shown by the female'. More recently, she has shown that pregnancy block does not occur when the female, the stud male and the strange male are all of the same inbred strain.[51]

In a review of 'smell as an exteroceptive factor' Bruce[50] concluded that 'the block to implantation depends upon the inhibition of prolactin secretion' (see above) 'while the effect of the male on oestrus depends on the evocation of FSH secretion'. Since the hypothalamic PIF has an inhibiting effect on prolactin secretion and FSH-RF has a stimulating effect on FSH secretion, it is conceivable that both effects are achieved by the same impulse transmitted from the outside world through the higher brain centres. Differences in the intensity of the hypothalamic response may, therefore, account for such differences as the occurrence of pseudopregnancy (Lee and Boot) or of di-oestrus (Whitten) in response to rather similar experimental manipulation.

Richmond and Conaway suggested that a variety of environmental factors have a non-specific, often a synchronizing, effect on the ovarian cycle, while the presence of the male may exert a more definite effect. It is possible, for example, that the oestrus-synchronizing effect of the introduction of a ram to a flock of ewes, described above, is a 'stress' effect in response to stimuli transmitted from the adrenal to the pituitary. The male was not involved in the synchronization of the onset of oestrus in gilts (nulliparous female pigs) apparently caused by their being transported from one area to another. An American team, wishing to study infertility and early embryonic mortality in pigs, collected a large number of sows from surrounding farms and brought them to the university farm. These animals had all been served at several successive periods of heat, but had failed to 'hold to service' and were therefore classed as 'repeat-breeders' (this is an example of the American use of 'to breed' meaning 'to mate'). The fact that a very high proportion of these pigs became pregnant at the first service on the university farm was apparently attributed to the 'natural cussedness' of things, but it may well be that a genuine physiological mechanism was involved.

An effect which was the reverse of that just suggested for pigs—namely, lowered fertility as the result of being moved to a strange environment—has been described in cattle. Cows at progeny-testing stations both in Scandinavia and in Britain were found to have a lower conception rate than those in farm conditions. The Dartington Hall Cattle Breeding Centre, in Devon, compared the conception rate (actually, the 'non-return' rate) of cows that had recently been moved to the farm when they were inseminated, with that of cows born on the farm. Fertility was significantly higher in the latter. The depression of fertility after moving lasted for 6 months or longer; this is well past the time when the urinary excretion of adreno-cortical steroid metabolites has returned to normal after an initial rise lasting 2–3 months.[199]

An interesting situation has recently been described in a population of the migratory wildebeeste of the Serengeti plain in East Africa.[386] These animals have a very restricted annual mating season, confined to a period of about 3 weeks early in the year. There is evidence that 'social' factors operate to synchronize oestrus among the cows. Few become pregnant at the first heat, but the conception rate at the second heat about 16 days later, is high. The great majority of cows experience two, and only two, periods of oestrus in each year.

Synchrony of oestrus in response to 'social' factors has also been demonstrated in non-murine rodents including the American deer mouse *Peromyscus maniculatus* and the prairie vole *Microtus ochrogaster*. The latter was the animal studied by Richmond and Conaway, in work already referred to.[311] They found that 'different manipulations in the conditions of housing would stimulate females to differing levels of reproductive activity, varying from simple opening of the vagina to persistent vaginal cornification and receptivity'.

The cottontail rabbit of America (*Sylvilagus floridanus*) and the European wild rabbit (*Oryctolagus cuniculus*) are seasonal breeders; many observers have noted the synchrony of the onset of oestrus throughout a population at the beginning of the breeding season. It has been suggested that social stimuli are important in inducing this synchrony, and it is clear that 'male-female social interactions' occur with increasing frequency just before breeding starts. In one prominent behaviour pattern the male repeatedly urinates in the direction of the female, and Marsden and Bronson, who investigated the pheromone effects of urine in mice, declared: 'It is a reasonable possibility that urine is the vehicle for a pheromone in the male rabbit which might aid in the induction of oestrus in the female, and that the mechanisms operating in *Sylvilagus* and *Oryctolagus* are similar to those in *Mus* and *Peromyscus*'.[243] These authors also suggested that high population density (or crowding) has an 'oestrus-suppressing effect', but this was questioned by Richmond and Conaway, who found that most 'disturbances' tended to induce oestrus in their colony of *Microtus*. Fighting is common among these animals and they found that 'severely bitten Microtus are nearly always found to be in oestrus'. They also quote the work of Petrusewicz who found a general increase of fertility in 47 'populations' of mice transferred from cage to cage, irrespective of the cage dimensions,[296] and that of Sheppe who reported 'unseasonal' breeding in colonies of deer mice from the vicinity of Lake Opinicon (Ontario) after their transfer to islands previously unoccupied by this species.[347] It seems very possible that both negative and positive effects of hypothalamic stimulation are involved.

These two series of experiments, in Poland and in Canada, form an interesting comparison. Both indicated an increase in prolificacy as a result of a change of environment, one in laboratory mice housed in cages, and the other in wild deer-mice unconfined on islands near their original habitat. Petrusewicz used cages of varied design and size. He was at pains to ensure that the population change that was observed after cage-change was not fortuitous, and in fact he was able to show that the eventual increase in numbers was greater than was to be expected had such factors not been involved. Sheppe, of course, had the residual 'mainland' population as a control. The population growth in the caged mice was due to an increase in the number of young that survived to breeding age as well as an increased breeding-rate. Petrusewicz stated that 'Most of the increase (i.e., in population number) occurred as the result of the survival of the unweaned mice born after the change of cage'. The increase in population number was usually preceded by a decline lasting several weeks, but this was due to increased mortality. The fertility of survivors increased to such an extent that the average number of young born in the month after cage-change was 20·7, as compared with 14·4 in the immediately preceding month.

It will be clear from what has been said that 'social' factors often appear to exert a 'stress' effect. In this context, 'stress' means a change in an animal's environment (external or internal) that demands some physiological adaptation. Adrenalectomized animals have little ability to respond to, or withstand, non-specific 'stressors' such as cold, infection, trauma, etc., but when an intact animal is subject to such stress, the pituitary-adrenal axis is activated and a fairly well defined sequence of reactions is called forth. The response was called the 'general adaptation syndrome' by Selye who, in 1946, put forward the hypothesis that a variety of pathological conditions may be due to malfunction of this response, or to the animal's increased vulnerability during the period of enhanced adreno-cortical activity.[343] He was able to induce diseases such as hypertension and arteriosclerosis by administering adrenal steroids to 'sensitized' animals. As the adrenal cortex secretes a variety of steroids, and as these affect the hypothalamus and hence the pituitary gland, as well as exerting their own direct effects, it is not surprising that adrenal activity may affect gonadal function. In unusual conditions or in some experimental situations, this interference may lead to reproductive malfunction, but in normal circumstances it may play an essential part in the reproductive rhythm. Reference is made elsewhere (p. 146) to the probable role of the foetal adrenal in parturition; our present concern is with changes in the oestrous cycle and the onset of the breeding season. Recent experiments have shown marked changes in gonadotrophin levels as a result of experimentally induced stress. For example, when mice of an inbred and non-aggressive strain were subjected to repeated 'aggression and defeat' (outright bullying in fact) by mice of a 'fighter' strain, there was a decided increase in the LH concentration in the pituitary gland and in the plasma of the bullied mice.[117] Morphological changes (including uterine hypoplasia) have been described in hamsters subjected to cold, or deprived of light,[310] and functional failure, including the inhibition of ovulation or implantation, may be caused by a period of fasting before oestrus.[235] There is a further complication in that the changes induced by keeping the animals in darkness were prevented if the pineal body was removed beforehand. The pineal body is known to affect reproductive function, probably by the secretion of methoxyindoles, exerting an inhibitory effect on the gonad (p. 158).

Red deer. Social factors play a prominent part in the modulation of reproductive physiology in species which exhibit a marked degree of agonistic, or aggressive, behaviour, particularly in the form of sexual competition between males. The red deer (*Cervus elaphus*) is perhaps the best known mammalian species of this kind, but it is only recently that behaviour has been related to hormonal status in individual deer, clarifying the real status of 'the monarch of the glen', his rivals and his harem. As in the case of other 'game' animals, a great deal of information is amassed by those engaged in this aspect, but even when it exists in recorded form, this lore is of little use to the physiologist, although his observations serve to explain the phenomena in retrospect.

The general behaviour of the red deer was vividly described, evidently from first-hand observation, by Turbervile in 1576,[372] and it was the subject of a pioneering study in mammalian ecology by Fraser Darling in 1937.[114] The red deer of the Scottish island of Rhum have recently been studied by Cambridge endocrinologists in collaboration with Edinburgh ecologists under the aegis of the Nature Conservancy.[224] In this work, all the stags in the study area were marked and, for record purposes, named (as a more attractive identification than a number). Two 'food points' were maintained by the experimenters to facilitate observation; the degree of 'interference' with the 'natural' habitat was judged to be slight, and acceptable in view of the observations made possible.

'Dominance' is important in the life of the red deer stag all the year

G

round, but for 9 or 10 months, from the end of November to about mid-September the stags live in all-male groups, covering a huge area of ground. Their year may be said to begin in April (Fig. 16), when the old antlers are cast and new ones immediately begin to grow. The gonads become active about the time of the summer solstice and the rising level of circulating testosterone causes the 'cleaning' of the antlers early in

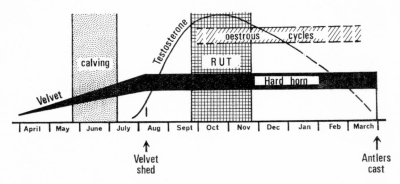

FIG. 16. Reproductive activity in the red deer.

The intensity of mating activity during rut results in most of the hinds becoming pregnant very early in their breeding season, and concentrates the period of calving in or near the following June.

August. The social hierarchy becomes very clearly defined while the stags are 'in hard horn', but as the testosterone level nears its maximum in September individual stags leave the male groups and seek out the hinds, who live in 'family groups' throughout the year. Calving is from late May to early July, and the females begin to experience a succession of oestrus cycles, of about 18 days, at the beginning of October after the beginning of the rut. These continue for about 6 months, in animals that do not become pregnant, and the stags are capable of service throughout this time. Most hinds are served and become pregnant during the 'rut', the period of concentrated sexual activity when the individual stags appropriate a group of hinds. During this time dominance is expressed as sexual success, but the hierarchy of the all-male groups is not necessarily maintained. It is not entirely clear what brings about the end of the rut after only a few weeks of intense sexual activity, while both sexes remain capable of breeding for months afterwards.

One might suppose the males to be exhausted, having had virtually no time or inclination to feed during the rut, but when at the end of it they re-form in all-male groups, the social hierarchy is re-established and fighting and other aggressive behaviour is at its most intense.

This complex behavioural pattern serves to concentrate the actual breeding season, but any females that have not become pregnant during the brief period of the rut are almost certain to be served in the succeeding months. Calving is thus concentrated during a favourable time of year, but early reproductive failures are not lost to the breeding potential.

References

An asterisk denotes a review article,
an outstanding contribution or a useful
guide to further reading.

1. ACKER, G. and ALLOITEAU, J. J. 1968. Le tout jeune corps jaune cyclique de la ratte est-il fonctionnel sans soutien hypophysaire? *C.r. Séanc. Soc. Biol.* **162**, 29.

2. ADAMS, C. E. 1965. Influence of the number of corpora lutea on endometrial proliferation and embryo development in the rabbit. *J. Endocr.* **31**, xxix.

*3. ALLEN, E. 1922. The oestrous cycle in the mouse. *Am. J. Anat.* **30**, 297.

4. ALLEN, W. M. and CORNER, G. W. 1930. Physiology of the corpus luteum VII. Maintenance of pregnancy in rabbit after very early castration, by corpus luteum extracts. *Proc. Soc. exp. Biol. Med.* **27**, 403.

5. ALLEN, W. R. 1969. The immunological measurement of pregnant mare serum gonadotrophin. *J. Endocr.* **43**, 593.

*6. AMOROSO, E. C. 1952. Placentation. In: *Marshall's Physiology of reproduction.* Ed. A. S. Parkes, Vol. 2, p. 127. Longmans, London.

7. — 1955. Endocrinology of pregnancy. *Brit. med. Bull.* **11**, 117.

*8. — and FINN, C. A. 1962. Ovarian activity during gestation, ovum transport and implantation. In: *The ovary.* Ed. S. Zuckerman, Vol. 1, p. 451. Academic Press, New York.

9. — HANCOCK, J. L. and ROWLANDS, I. W. 1948. Ovarian activity in the pregnant mare. *Nature, Lond.* **161**, 355.

10. — and ROWLANDS, I. W. 1951. Hormonal effects in the pregnant mare and foetal foal. *J. Endocr.* **7**, 1.

11. ANAND KUMAR, T. C. 1968. Oogenesis in lorises; *Loris tardigradus* and *Nycticebus coucang. Proc. R. Soc.* **169**, 167.

12. ANDERSON, L. L., BOWERMAN, A. M. and MELAMPY, R. M. 1963. Neuro-utero-ovarian relationships. In: *Advances in neuroendocrinology.* Ed. A. V. Nalbandov, p. 345. University of Illinois Press, Urbana, Illinois.

13. — DYCK, G. W., MORI, H., HENRICKS, D. and MELAMPY, R. M. 1967. Ovarian function in pigs following hypophysial stalk transection or hypophysectomy. *Am. J. Physiol.* **212**, 1188.

14. — RATHMACHER, R. P. and MELAMPY, R. M. 1966. The uterus and unilateral regression of corpora lutea in the pig. *Am. J. Physiol.* **210**, 611.

*15. APOSTOLAKIS, M. 1968. Prolactin. *Vitams Horm.* **26**, 197.

16. ARMSTRONG, D. T. 1968. *In vitro* synthesis of progesterone. *J. Anim. Sci.* **27**, Suppl., 181.

17. Aron, C. 1965. Données nouvelles sur les mécanismes de la ponte chez les femelles dites 'à ponte spontanée'. Mise en évidence, chez la ratte, de l'action ovulatoire du coït au cours du cycle oestral. *Revue Roumaine d'Endocr.* **2,** 221.

18. — Asch, G. and Roos, J. 1965. Mise en évidence de l'action ovulatoire du rapprochement sexuel chez la ratte gestante. *C.r. Séanc. Soc. Biol.* **159,** 2505.

19. — Roos, J. and Asch, G. 1967. Données nouvelles sur le rôle joué par la ponte provoquée dans les phenomènes de la reproduction chez la ratte. *Annls. Endocr.* **28,** 19.

*20. Asdell, S. A. 1964. *Patterns of mammalian reproduction. 2nd edition.* Constable, London.

21. Astwood, E. B. 1941. The regulation of corpus luteum function by hypophysial luteotrophin. *Endocrinology* **28,** 309.

22. — and Fevold, H. L. 1939. Action of progesterone on the gonadotropic activity of the pituitary. *Am. J. Physiol.* **127,** 192.

23. Baca, M. and Zamboni, L. 1967. Fine structure of human follicular oocytes. *J. ultrast. Res.* **19,** 354.

24. Banon, P., Brandes, D. and Frost, J. K. 1964. Lysosomal enzymes in the rat ovary and endometrium during the estrous cycle. *Acta cytol.* **8,** 416.

25. Barbanti Silva, C., Albertazzi, E., Trentini, G. P. and Botticelli, A. 1966. Modificazioni istomorfologiche ed istoenzimatiche dell'ovaio e dell'utero di ratta adulta conseguenti ad epifisectomia. *Riv. Ital. Ginec.* **50,** 515.

26. Barnett, S. A. and Mount, L. E. 1967. Resistance to cold in mammals. In: *Thermobiology.* Ed. A. H. Rose. Academic Press, New York.

*27. Barraclough, C. A. and Leathem, J. H. 1954. Infertility induced in mice by a single injection of testosterone propionate. *Proc. Soc. exp. Biol. Med.* **85,** 673.

28. Bjersing, L. 1967. Ultrastructure of granulosa lutein cells in porcine corpus luteum, with special reference to endoplasmic reticulum and steroid hormone synthesis. *Z. Zellforsch. mikrosk. Anat.* **82,** 187.

29. Black, J. L. and Erickson, B. H. 1968. Oogenesis and ovarian development in the prenatal pig. *Anat. Rec.* **161,** 45.

30. Blanchette, E. Joan 1966. Ovarian steroid cells: 1. Differentiation of the lutein cell from the granulosa follicle cell during the pre-ovulatory stage and under the influence of exogenous gonadotrophins. 2. The lutein cell. *J. Cell Biol.* **31,** 501, 517.

31. Blandau, R. J. 1965. Observations on the migration of living primordial germ cells in the mouse. *Anat. Rec.* **151,** 498.

32. Bodenheimer, F. S. 1949. Problems of vole populations in the Middle East. Research Council of Israel, Azriel Printing Works, Jerusalem.

33. BOOKHOUT, C. G. 1945. The development of the guinea-pig ovary from sexual differentiation to maturity. *J. Morph.* **77**, 233.

*34. BOUIN, P. 1902. Les deux glandes à sécrétion interne de l'ovaire; la glande interstitielle et le corps jaune. *Rev. Med. Est.* **34**, 465.

35. — and ANCEL, P. 1909. Sur la fonction du corps jaune. Action du corps jaune vrai sur l'utérus. *C.r. Séanc. Soc. Biol.* **66**, 505.

36. — and ANCEL, P. 1909. Sur la fonction du corps jaune. Démonstration expérimentale de l'action du corps jaune sur l'utérus et la glande mammaire. *C.r. Séanc. Soc. Biol.* **66**, 689.

37. BRAMBELL, F. W. R. 1927. The development and morphology of the gonads of the mouse. 1. The morphogenesis of the indifferent gonad and of the ovary. *Proc. R. Soc., B.* **101**, 391.

38. — 1930. *The development of sex in vertebrates.* Sidgwick & Jackson, London.

39. — 1937. The influence of lactation on the implantation of the mammalian embryo. *Am. J. Obstet. Gynec.* **33**, 942.

40. — 1944. The reproduction of the wild rabbit, *Oryctolagus cuniculus* (L.) *Proc. zool. Soc. Lond.* **114**, 1.

*41. — 1956. Ovarian changes. In: *Marshall's Physiology of reproduction*, 3rd Ed. Ed. A. S. Parkes, Vol. 1, Pt. 1, p. 397.

42. — and HALL, K. 1939. Reproduction of the field vole, *Microtus agrestis hirtus* Bellamy. *Proc. zool. Soc. Lond., A.* **109**, 133.

43. — and ROWLANDS, I. W. 1936. Reproduction of the bank vole (*Evotomys glareolus*, Schreber). *Phil. Trans. R. Soc. Ser. B* **226**, 71.

44. BREED, W. G. 1969. Oestrus and ovarian histology in the lactating vole (*Microtus agrestis*). *J. Reprod. Fert.* **18**, 33.

45. BRETSCHNEIDER, L. H. and DUYVENÉ DE WIT, J. J. 1947. *Sexual endocrinology of non-mammalian vertebrates.* Elsevier, Amsterdam.

*46. BRODISH, A. 1968. A review of neuroendocrinology. *Yale J. Biol. Med.* **41**, 143.

47. BRONSON, F. H. and WHITTEN, W. K. 1968. Oestrus-accelerating pheromone of mice; assay, androgen-dependency and presence in bladder urine. *J. Reprod. Fert.* **15**, 131.

48. BROWN-GRANT, K. 1966. The action of hormones on the hypothalamus. *Br. med. Bull.* **22**, 273.

49. BRUCE, H. M. 1960. A block to pregnancy in the mouse caused by the proximity of strange males. *J. Reprod. Fert.* **1**, 96.

*50. — 1966. Smell as an exteroceptive factor. *J. Anim. Sci.* **25**, Suppl., 83.

51. — 1968. Absence of pregnancy-block in mice when stud and test males belong to an inbred strain. *J. Reprod. Fert.* **17**, 407.

52. — and HINDLE, E. 1934. The golden hamster, *Cricetus* (*Mesocricetus*) *auratus* Waterhouse. Notes on its breeding and growth. *Proc. zool. Soc. Lond.* **104**, 361.

53. — RENWICK, A. G. C. and FINN, C. A. 1968. Effect of post-coital unilateral ovariectomy on implantation in mice. *Nature, Lond.* **219,** 733.

54. BUCHANAN, G. D., ENDERS, A. C. and TALMAGE, R. V. 1956. Implantation in armadillos ovariectomized during the period of delayed implantation. *J. Endocr.* **14,** 121.

55. BUECHNER, H. K., MORRISON, J. A. and LEUTHOLD, W. 1966. Reproduction in Uganda kob with special reference to behavior. In: *Comparative biology of reproduction in mammals.* Ed. I. W. Rowlands, p. 69. Academic Press, New York.

56. — and SWANSON, C. V. 1955. Increased natality resulting from lowered population density among elk in southwestern Washington. *Trans. N. Am. Wildl. Conf.* **20,** 560.

57. BURGER, J. F. 1952. Sex physiology of pigs. *Onderstepoort J. vet. Res.* **25,** Suppl. 2.

58. BUTLER, H. 1967. The oestrous cycle of the Senegal bush baby in the Sudan. *J. Zool.* **151,** 143.

*59. BUTT, W. R. 1967. *Hormone Chemistry.* Van Nostrand, London.

60. CALVIN, G. B. and SAWYER, C. H. 1969. Induction of running behavior in ovariectomized rats by implanting estrogen into the hypothalamus. *Anat. Rec.* **163,** 171.

61. CAMPBELL, H. J. and GALLARDO, E. 1966. Gonadotrophin-releasing activity of the median eminence at different ages. *J. Physiol., Lond.* **186,** 689.

62. CANIVENC, R. and BONNIN-LAFFARGUE, M. 1963. Inventory of problems raised by the delayed ova implantation in the European badger (*Meles meles L.*). In: *Delayed Implantation.* Ed. A. C. Enders, University of Chicago Press.

63. — SHORT, R. V. and BONNIN-LAFFARGUE, M. 1966. Étude histologique et biochemique du corps jaune du blaireau européen (*Meles meles* L.). *Annls Endocr.* **27,** 401.

64. CATCHPOLE, H. R. and COLE, H. H. 1934. The distribution and source of oestrin in the pregnant mare. *Anat. Rec.* **59,** 335.

65. CHAN, S. T. H., WRIGHT, A. and PHILLIPS, J. G. 1967. The atretic structures in the gonads of the rice-field eel (*Monopterus albus*) during natural sex-reversal. *J. Zool.* **153,** 527.

66. CHANG, M. C. 1955. The maturation of rabbit oocytes in culture and their maturation, activation, fertilization and subsequent development in the fallopian tube. *J. exp. Zool.* **128,** 379.

67. CHAPLIN, R. E. 1966. *Reproduction in British deer.* Passmore Edwards Museum, London.

68. CHIPMAN, R. K. and FOX, K. A. 1966. Oestrus synchronization and pregnancy blocking in wild house mice (*Mus musculus*). *J. Reprod. Fert.* **12,** 233.

69. CHITTY, D. 1952. Mortality among voles (*Microtus agrestis*) at Lake Vyrnwy, Montgomeryshire in 1936–9. *Phil. Trans. R. Soc., Ser. B.* **236**, 505.

70. CHRISTIAN, J. J. 1950. The adreno-pituitary system and population cycles in mammals. *J. Mammal.* **31**, 247.

71. CLARKE, J. R. and KENNEDY, J. P. 1967. Changes in the hypothalamo-hypophysial neurosecretory system and the gonads of the vole (*Microtus agrestis*). *Gen. comp. Endocr.* **8**, 455.

72. COLE, H. H. and HART, G. H. 1930. The potency of blood serum of mares in progressive stages of pregnancy in affecting the sexual maturity of the immature rat. *Am. J. Physiol.* **93**, 57.

73. — HART, G. H., LYONS, W. R. and CATCHPOLE, H. R. 1933. The development and hormonal content of foetal horse gonads. *Anat. Rec.* **56**, 275.

*74. — HOWELL, C. E. and HART, G. H. 1931. The changes occurring in the ovary of the mare during pregnancy. *Anat. Rec.* **49**, 199.

75. COMLINE, R. S. and SILVER, M. 1961. The release of adrenaline and noradrenaline from the adrenal glands of the foetal sheep. *J. Physiol., Lond.* **156**, 424.

76. CONAWAY, C. H. 1959. The reproductive cycle of the eastern mole. *J. Mammal.* **40**, 180.

77. CONAWAY, C. H. and SORENSON, M. W. 1966. Reproduction in tree shrews. In: *Comparative biology of reproduction in mammals*. Ed. I. W. Rowlands, p. 471. Academic Press, New York.

78. COOPER, E. and HESS, M. 1968. Demonstration of a utero-ovarian relationship independent of species. *Anat. Rec.* **160**, 335.

*79. CORNER, G. W. 1928. Physiology of the corpus luteum. 1. The effect of very early ablation of the corpus luteum upon embryos and uterus. *Am. J. Physiol.* **86**, 74.

80. — 1938. The sites of formation of estrogenic substances in the animal body. *Physiol. Rev.* **18**, 154.

81. COUTTS, R. R. and ROWLANDS, I. W. 1969. The reproductive cycle of the Skomer vole (*Clethrionomys glareolus skomerensis*). *J. Zool., Lond.* **158**, 1.

82. CRAIGHEAD, J. J., HORNOCKER, M. G. and CRAIGHEAD, F. C., Jr. 1969. Reproductive biology of young female grizzly bears. *J. Reprod. Fert.* Suppl. **6**, 447.

*83. CSAPO, A. 1956. Progesterone "block". *Am. J. Anat.* **98**, 273.

84. — 1969. The four direct regulatory factors of myometrial function. In: *Progesterone: its regulatory effect on the myometrium*. (Ciba Foundation Study Group No. 34). Ed. G. E. W. Wolstenholme and Julie Knight, p. 13. Churchill, London.

85. DANIEL, M. J. 1963. Early fertility of red deer hinds in New Zealand. *Nature, Lond.* **200**, 380.

*86. DANIEL, P. M. 1966. The blood supply of the hypothalamus and pituitary gland. *Brit. med. Bull.* **22**, 202.

87. DAUZIER, L., ORTAVANT, R., THIBAULT, C. and WINTEÑBERGER, S. 1953. Recherches expérimentales sur le role de la progestérone dans le cycle sexuel de la brebis et de la chèvre. *Annls. Endocr.* **14**, 553.

88. DAVID, M. A., FRASCHINI, F. and MARTINI, L. 1966. Control of LH secretion: role of a "short" feedback mechanism. *Endocrinology* **78**, 55.

89. DEANE, HELEN W. 1958. Intracellular lipides: their detection and significance. In: *Frontiers in cytology.* Ed. S. L. Palay, p. 227. Yale University Press, New Haven, Connecticut.

90. — HAY, MARY F., MOOR, R. M., ROWSON, L. E. A. and SHORT, R. V. 1966. Corpus luteum of the sheep: relationships between morphology and function during the oestrous cycle. *Acta Endocr., Copnh.* **51**, 245.

91. DEANESLY, RUTH 1934. The reproductive processes of certain mammals. Part VI—The reproductive cycle of the female hedgehog. *Phil. Trans. R. Soc. Ser. B.* **223**, 239.

92. — 1938. The reproductive cycle of the golden hamster (*Cricetus auratus*). *Proc. zool. Soc. Lond., A.* **108**, 31.

93. — 1944. The reproductive cycle of the female weasel (*Mustela nivalis*). *Proc. zool. Soc. Lond.* **114**, 339.

94. — 1963. The corpus luteum hormone during and after ovo-implantation: an experimental study of its mode of action in the guinea-pig. In: *Delayed implantation.* Ed. A. C. Enders. University of Chicago Press.

*95. — 1966a. The endocrinology of pregnancy and foetal life. In: *Marshall's Physiology of reproduction,* 3rd edition. Ed. A. S. Parkes, Vol. 3, p. 891. Longmans, London.

96. — 1966b. Observations on reproduction in the mole *Talpa europaea.* In: *Comparative biology of reproduction in mammals.* Ed. I. W. Rowlands, p. 387. Academic Press, New York.

97. DE DUVE, C. 1964. From cytases to lysosomes. *Fed. Proc.* **23**, 1045.

98. DE JONGH, S. E. and WOLTHUIS, O. L. 1964. Factors determining cessation of corpus luteum function; possible role of oestradiol and progesterone. *Acta endocr., Copnh.* **45**, Suppl. 90, 125.

99. Dempsey, E. W. 1937. Follicular growth rate and ovulation after various experimental procedures in the guinea-pig. *Am. J. Physiol.,* **120**, 126.

100. DENAMUR, R., MARTINET, J. and SHORT, R. V. 1966. Progesterone secretion by the sheep corpus luteum after hypophysectomy, pituitary stalk section and hysterectomy. *Acta endocr., Copnh.* **52**, 72.

101. — and MAULEÓN, P. 1963. Effets de l'hypophysectomie sur la morphologie et l'histologie du corps jaune des ovins. *C.r. hebd. Séanc. Acad. Sci., Paris.* **257**, 264.

102. DINGLE, J. T., HAY, MARY F. and MOOR, R. M. 1968. Lysosomal function in the corpus luteum of the sheep. *J. Endocr.* **40**, 325.

194 THE OVARIAN CYCLE OF MAMMALS

103. DÖCKE, F. and DÖRNER, G. 1969. A possible mechanism by which progesterone facilitates ovulation in the rat. *Neuroendocrinology* **4**, 139.

*104. DONALDSON, H. H. 1924. The rat: data and reference tables. *Mem. Wistar Inst. Anat. Biol.*, No. 6, 2nd edition.

105. DONALDSON, L. and HANSEL, W. 1965. Histological study of bovine corpora lutea. *J. Dairy Sci.* **48**, 905.

106. DONOVAN, B. T. 1967. Control of follicular growth and ovulation. In: *Reproduction in the female mammal.* Ed. G. E. Lamming and E. C. Amoroso. Butterworth, London.

*107. — and HARRIS, G. W. 1966. Neurohumoral mechanisms in reproduction. In: *Marshall's Physiology of reproduction.* 3rd ed. Ed. A. S. Parkes, vol. 3, p. 301. Longmans, London.

108. — and O'KEEFE, MARY C. 1966. The liver and the feedback action of ovarian hormones in the immature rat. *J. Endocr.* **34**, 469.

109. DRESEL, I. 1935. The effect of prolactin on the estrus cycle of non-parous mice. *Science* **82**, 173.

110. DUBREUIL, G. 1957. Le déterminisme de la glande thécale de l'ovaire. Induction morphogène à partir de la granulosa folliculaire. *Acta anat.* **30**, 269.

111. DU MESNIL DU BUISSON, F. 1961a. Régression unilatérale des corps jaunes après hystérectomie partielle chez la truie. *Annls Biol. anim. Biochim. Biophys.* **1**, 105.

*112. — 1961b. Possibilité d'un fonctionnement dissemblable des ovaires pendant la gestation chez la truie. *C.r. hebd. Séanc. Acad. Sci., Paris* **253**, 727.

113. — and LÉGLISE, P. C. 1963. Effet de l'hypophysectomie sur les corps jaunes de la truie. Résultats préliminaires. *C.r. hebd. Séanc. Acad. Sci., Paris* **257**, 261.

114. ECKSTEIN, P. and ZUCKERMAN, S. 1956. Morphology of the reproductive tract. In: *Marshall's Physiology of reproduction*, 3rd Edition. Ed. A. S. Parkes, Vol. 1, Pt. 1, 43. Longmans, London.

*115. — and ZUCKERMAN, S. 1956. The oestrous cycle in the mammalia. In: *Marshall's Physiology of reproduction*, 3rd Edition. Ed. by A. S. Parkes, Vol. 1, Pt. 1, p. 226. Longmans, London.

116. EDWARDS, R. G. 1962. Meiosis in ovarian oocytes of adult mammals. *Nature, Lond.* **196**, 446.

117. ELEFTHERIOU, B. E. and CHURCH, R. L. 1967. Effects of repeated exposure to aggression and defeat on plasma and pituitary levels of luteinizing hormone in C57 BL/6J mice. *Gen. comp. Endocr.* **9**, 263.

*118. ENDERS, A. C. (Ed.) 1963. *Delayed implantation.* University of Chicago Press.

119. ENDERS, R. K. 1952. Reproduction in the mink (*Mustela vison*). *Proc. Am. phil. Soc.* **96**, 691.

120. ENGLAND, B. G., FOOTE, W. C., MATTHEWS, D. H., CARDOZO, A. G. and RIERA, S. 1969. Ovulation and corpus luteum function in the llama (*Lama glama*). *J. Endocr.* **45**, 505.

121. ERICKSON, B. H. 1966. Development and radio-response of the prenatal bovine ovary. *J. Reprod. Fert.* **11**, 97.

122. ESPEY, L. L. 1967. Ultrastructure of the apex of the rabbit graafian follicle during the ovulatory process. *Endocrinology* **81**, 267.

123. EVANS, H. M., SIMPSON, M. E., LYONS, R. and TURPEINEN, K. 1941. Anterior pituitary hormones which favor the production of traumatic uterine placentomata. *Endocrinology* **28**, 933.

124. EVERETT, J. W. 1948. Progesterone and estrogen in the experimental control of ovulation time and other features of the estrous cycle in the rat. *Endocrinology* **43**, 389.

125. — 1956. Functional corpora lutea maintained for months by autografts of rat hypophyses. *Endocrinology* **58**, 786.

126. — 1956. The time of release of ovulating hormone from the rat hypophysis. *Endocrinology* **59**, 580.

127. EWART, J. C. 1897. *A critical period in the development of the horse.* Adam & Charles Black, London.

128. FALCK, B. 1959. Site of production of oestrogen in rat ovary as studied in micro-transplants. *Acta physiol. scand.* **47**, Suppl., 163.

*129. FARRIS, E. J. and GRIFFITH, J. Q., Jr. (Eds.) 1949. *The rat in laboratory investigation.* 2nd ed. Lippincott, Philadelphia.

*130. FEE, A. R. and PARKES, A. S. 1929. Studies on ovulation. 1. The relation of the anterior pituitary body to ovulation in the rabbit. *J. Physiol.* **67**, 383.

131. FISCHER, T. V. 1967. Local uterine regulation of the corpus luteum. *Am. J. Anat.* **121**, 425.

132. FLETCHER, I. C. and LINDSAY, D. R. 1968. Sensory involvement in the mating behaviour of domestic sheep. *Anim. Behav.* **16**, 410.

133. FLUX, J. E. C. 1965. Timing of the breeding season in the hare, *Lepus europaeus*, and rabbit, *Oryctolagus cuniculus*. *Mammalia* **29**, 557.

134. FLYNN, T. T. and HILL, J. P. 1939. The development of the Monotremata. Part IV. Growth of the ovarian ovum, maturation, fertilisation, and early cleavage. *Trans. zool. Soc. Lond.* **24**, 445.

135. FOOTE, W. C., ENGLAND, B. G. and WILDE, M. E. 1968. Llama reproduction—a South American problem. *Utah Sci.* **29**, 43.

136. FORD, C. S. and BEACH, F. A. 1952. *Patterns of Sexual Behaviour.* Eyre & Spottiswode, London.

137. FOSTER, M. A. 1934. The reproductive cycle in the female ground-squirrel (*Citellus tridecemlineatus* M.). *Am. J. Anat.* **54**, 487.

*138. FRAENKEL, L. 1910. Neue Experimente zur Function des Corpus Luteum. *Arch. Gynäk.* **91**, 705.

*139. FRANCHI, L. L., MANDL, ANITA M. and ZUCKERMAN, S. 1962. The development of the ovary and the process of oogenesis. In: *The ovary*. Vol. I, 1. Ed. S. Zuckerman. Academic Press, New York.

140. FRAPS, R. M. 1962. Effects of external factors on the activity of the ovary. In: *The ovary*: vol. 2. Ed. S. Zuckerman. Academic Press, New York.

141. FRASER DARLING, F. 1937. *A herd of red deer*. Oxford University Press, London.

*142. GALA, R. R. and WESTPHAL, U. 1967. Corticosteroid binding activity in serum of mouse, rabbit and guinea-pig during pregnancy and lactation: possible involvement in the initiation of lactation. *Acta Endocr. Copnh.* **55,** 47.

143. GALIL, A. K. A. 1965. Utero-ovarian interrelations during pregnancy— endocrine role of the placenta. Ph.D. Thesis, London University.

144. — and DEANE, HELEN W. 1966. Δ^5-3β-Hydroxysteroid dehydrogenase activity in the steroid-hormone producing organs of the ferret (*Mustela putorius furo*). *J. Reprod. Fert.* **11,** 333.

145. GARDE, M. L. 1930. The ovary of *Ornithorhynchus*, with special reference to follicular atresia. *J. Anat., Lond.* **64,** 422.

146. GLENISTER, T. W. 1963. Observations on mammalian blastocysts implanting in organ culture. In: *Delayed implantation*. Ed. A. C. Enders, p. 171. University of Chicago Press.

147. GRANT, R. 1933. Occurrence of ovulation without heat in ewe. *Nature, Lond.*, **131,** 802.

148. GREEN, J. A., GARCILAZO, J. A. and MAQUEO, M. 1967. Ultrastructure of the human ovary. 2. Luteal cell at term. *Am. J. Obstet. Gynec.* **99,** 855.

149. GREENWALD, G. S. 1957. Reproduction in a coastal California population of the field mouse, *Microtus californicus. Univ. Calif. Publs Zool.* **54,** 421.

150. GROTA, L. J. and EIK-NES, K. B. 1967. Plasma progesterone concentrations during pregnancy and lactation in the rat. *J. Reprod. Fert.* **13,** 83.

151. GUILLEMIN, R. 1968. Control of the secretions of the adenohypophysis by the central nervous system. In: *Proc. 24th Int. Congr. physiol. Sci.* (*Abstr.*) Vol. 6, p. 69.

152. HACKETT, A. J. and HAFS, H. D. 1969. Pituitary and hypothalamic endocrine changes during the bovine estrous cycle. *J. Anim. Sci.* **28,** 531.

153. HAFEZ, E. S. E. 1952. Studies on the breeding season and reproduction of the ewe. *J. agric. Sci.* **42,** 189.

*154. — (Ed.) 1962. *The behaviour of domestic animals*. Baillière, Tindall & Cox, London.

155. HALL, KATHLEEN 1957. The effect of relaxin extracts, progesterone and oestradiol on maintenance of pregnancy, parturition and rearing of young after ovariectomy in mice. *J. Endocr.* **15,** 108.

156. — 1960. Relaxin. *J. Reprod. Fert.* **1,** 368.

157. HAMERTON, J. L., DICKSON, JANET M., POLLARD, CAROLYN E., GRIEVES, SUSAN A. and SHORT, R. V. 1969. Genetic intersexuality in goats. *J. Reprod. Fert.*, Suppl. **7,** 25.

*158. HAMMOND, J. 1925. *Reproduction in the rabbit.* Oliver & Boyd, Edinburgh.

*159. — 1927. *The physiology of reproduction in the cow.* Cambridge University Press, London.

160. — 1960. *Farm animals*, 3rd edition. Edward Arnold, London.

161. HAMMOND, J., Jr. 1951. Failure of progesterone treatment to affect delayed implantation in mink. *J. Endocr.* **7,** 330.

162. — 1953. *Effects of artificial lighting on the reproductive and pelt cycles of mink.* Heffer, Cambridge.

163. — 1964. *The breeding season of the female ferret: on natural lighting and on days of constant length and intensity.* Heffer, Cambridge.

164. HANCOCK, J. L. and ROWLANDS, I. W. 1949. The physiology of reproduction in the dog. *Vet. Rec.* **61,** 771.

*165. HANSEL, W. and SEIFART, K. H. 1967. Maintenance of luteal function in the cow. *J. Dairy Sci.* **50,** 1948.

166. HANSSON, A. 1947. The physiology of reproduction in the mink (*Mustela vison*, Schreb.) with special reference to delayed implantation. *Acta zool., Stockh.* **28,** 1.

167. HARDISTY, M. W. 1967. The numbers of vertebrate primordial germ cells. *Biol. Rev.* **42,** 265.

168. HARDY, B. and LODGE, G. A. 1969. The influence of nutrition during post-lactational oestrus on ovulation rate in the sow and the accuracy of corpora lutea counts in estimating ovulations. *J. Reprod. Fert.* **19,** 555.

169. HARRIS, G. W. 1937. The induction of ovulation in the rabbit, by electrical stimulation of the hypothalamo-hypophysial mechanism. *Proc. R. Soc., B.* **122,** 374.

*170. — 1955. *Neural control of the pituitary gland.* Edward Arnold, London.

171. — and JACOBSOHN, D. 1952. Functional grafts of the anterior pituitary gland. *Proc. R. Soc. B.* **139,** 263.

172. HARRISON, R. J. 1962. The structure of the ovary: C. Mammals. In: *The ovary.* Vol. 1, p. 143. Ed. S. Zuckerman. Academic Press, New York.

173. — 1963. A comparison of factors involved in delayed implantation in badgers and seals in Great Britain. In: *Delayed implantation.* Ed. A. C. Enders. University of Chicago Press.

*174. HART, G. H. and COLE, H. H. 1934. The source of oestrin in the pregnant mare. *Am. J. Physiol.* **109,** 320.

175. HEAP, R. B. 1969. The binding of plasma progesterone in pregnancy. *J. Reprod. Fert.* **18,** 546.

176. — and DEANESLY, R. 1966. Progesterone in systemic blood and placentae of intact and ovariectomized pregnant guinea-pigs. *J. Endocr.* **34,** 417.

177. — and DEANESLY, R. 1967. The increase in plasma progesterone levels in the pregnant guinea-pig and its possible significance. *J. Reprod. Fert.* **14,** 339.

178. — and LINZELL, J. L. 1966. Arterial concentration, ovarian secretion and mammary uptake of progesterone in goats during the reproductive cycle. *J. Endocr.* **36,** 389.

179. — PERRY, J. S. and ROWLANDS, I. W. 1967. Corpus luteum function in the guinea-pig; arterial and luteal progesterone levels, and the effects of hysterectomy and hypophysectomy. *J. Reprod. Fert.* **13,** 537.

180. HEAPE, W. 1905. Ovulation and degeneration of ova in the rabbit. *Proc. R. Soc., B.* **76,** 260.

181. HECHTER, O., FRAENKEL, M., LEV, M. and SOSKIN, S. 1940. Influence of the uterus on the corpus luteum. *Endocrinology* **26,** 680.

182. HECKEL, G. P. 1942. The estrogen sparing effect of hysterectomy. *Surgery Gynec. Obstet.* **75,** 379.

183. HEDIGER, H. 1952. Unsere Zebras. *Die Umschau in Wissenschaft und Technik* **52,** 398.

184. HERBERT, J. 1970. Reproductive behaviour of rhesus and talapoin monkeys. *J. Reprod. Fert.*, Suppl. 11, 119.

185. HILL, C. J. 1941. The development of the Monotremata. Part V. Further observations on the histology and the secretory activities of the oviduct prior to and during gestation. *Trans. zool. Soc. Lond.* **25,** 1.

186. HILL, J. P. 1933. The development of the Monotremata. Part II. The structure of the egg-shell. *Trans. zool. Soc. Lond.* **21,** 443.

187. — and GATENBY, J. B. 1926. The corpus luteum of the Monotremata. *Proc. zool. Soc. Lond.* **47,** 715.

188. — and HILL, W. C. O. 1955. The growth-stages of the pouch-young of the native cat (*Dasyurus viverrinus*) together with observations on the anatomy of the new-born young. *Trans. zool. Soc. Lond.* **28,** 349.

189. HILL, M. and WHITE, W. E. 1934. The growth and regression of follicles in the oestrous rabbit. *J. Physiol.* **80,** 174.

190. HILL, W. C. O. 1953. *Primates: comparative anatomy and taxonomy.* Vol. 1. Wiley (Interscience), New York.

191. HILLIARD, J., ARCHIBALD, D. and SAWYER, C. H. 1963. Gonadotropic activation of preovulatory synthesis and release of progestin in the rabbit. *Endocrinology* **72,** 59.

192. — and SAWYER, C. H. 1964. Synthesis and release of progestin by rabbit ovary *in vivo*. In: *Hormonal steroids: biochemistry, pharmacology and therapeutics.* Vol. 1, p. 263 (Proc. 1st Int. Congr. Hormonal Steroids). Ed. L. Martini and A. Pecile. Academic Press, New York.

193. HINSEY, J. C. and MARKEE, J. E. 1933. Pregnancy following bilateral section of the cervical sympathetic trunks in the rabbit. *Proc. Soc. exp. Biol. Med.* **31,** 270.

194. HISAW, F. L. 1961. In: *Control of ovulation*. Ed. C. A. Villee. Harvard University, Cambridge, Massachusetts.

195. — HISAW, F. L., Jr. and DAWSON, A. B. 1967. Effects of relaxin on the endothelium of endometrial blood vessels in monkeys. *Endocrinology* **81**, 375.

196. HOAR, W. S. 1955. Reproduction in teleost fish. *Mem. Soc. Endocr.* **4**, 5.

197. HOFFMANN, J. C. 1967. Effects of light deprivation on the rat estrous cycle. *Neuroendocrinology* **2**, 1.

198. — and SCHWARTZ, N. B. 1964. 'Progesterone-withdrawal' ovulation in rat. *Fedn Proc. Fedn Am. Socs exp. Biol.* **23**, 109.

199. HOLT, A. F. 1962. Movement of cattle and its effect on fertility. *Brit. vet. J.* **118**, 293.

200. HUMPHREY, R. R. 1948. Reversal of sex in females of genotype WW in the axolotl (*Siredon* or *Ambystoma mexicanum*) and its bearing upon the role of the Z chromosomes in the development of the testis. *J. exp. Zool.* **109**, 171.

201. HUNTER, G. L. and LISHMAN, A. W. 1967. Effect of the ram early in the breeding season on the incidence of ovulation and oestrus in sheep. *Proc. S. Afr. Soc. Anim. Prod.* **6**, 199.

*202. HUNTER, J. 1787. An experiment to determine the effect of extirpating one ovarium upon the number of young produced. *Phil. Trans. R. Soc. Ser. B.* **77**, 233.

203. ILLINGWORTH, DOREEN V., HEAP, R. B. and PERRY, J. S. 1970. Changes in the metabolic clearance rate of progesterone in the guinea-pig. *J. Endocr.* **48**, 409.

*204. INGRAM, D. L. 1962. Atresia. In: *The ovary*. Ed. S. Zuckerman. Academic Press, New York.

205. IOANNOU, J. M. 1966. The oestrous cycle of the potto. *J. Reprod. Fert.* **11**, 455.

206. — 1967. Oogenesis in adult prosimians. *J. Embryol. exp. Morphol.* **17**, 139.

207. JARVIS, JENNIFER U. M. 1969. The breeding season and litter size of African mole-rats. *J. Reprod. Fert.*, Suppl. **6**, 237.

*208. JONES, ESTHER C. 1970. The ageing ovary and its influence on reproductive capacity. *J. Reprod. Fert.*, Suppl. **12**, 17.

209. JONES, ESTHER C. and KROHN, P. L. 1960. The effect of unilateral ovariectomy on the reproductive lifespan of mice. *J. Endocr.* **20**, 129.

210. JOST, A. 1955. Modalities in the action of gonadal and gonad-stimulating hormones in the foetus. *Mem. Soc. Endocr.* **4**, 237.

211. JOUBERT, D. M. 1963. Puberty in female farm animals. *Anim. Breed. Abstr.* **31**, 295.

212. KENT, G. C., Jr. 1968. Physiology of reproduction (of the golden hamster). In: *The golden hamster: its biology and use in medical research*. Ed. R. A. Hoffman, P. F. Robinson and Hulda Magalhaes. Iowa State University Press, Ames.

213. KIRBY, D. R. S. 1967. Ectopic autografts of blastocysts in mice maintained in delayed implantation. *J. Reprod. Fert.* **14**, 515.

214. LAMMING, G. E. 1966. Nutrition and the endocrine system. *Nutr. Abstr. Rev.* **36**, 1.

215. LANG, E. M. 1967. The birth of an African elephant, *Loxodonta africana*, at Basle zoo. *Int. Zoo Yb.* **7**, 154.

216. LATASTE, F. 1887. Notes prises au jour le jour sur différentes espèces de l'ordre des rongeurs observées en captivité. *Actes Soc. Linn. Bordeaux* **40**, 293.

217. LAURIE, E. M. O. 1946. The reproduction of the house-mouse (*Mus musculus*) living in different environments. *Proc. R. Soc., B.* **133**, 248.

218. LAWS, R. M. and CLOUGH, G. 1966. Observations on reproduction in the hippopotamus *Hippopotamus amphibius* Linn. In: *Comparative biology of reproduction in mammals*. Ed. I. W. Rowlands, p. 127. Academic Press, New York.

219. LEMAY, J. P. and CORRIVAULT, G. W. 1968. L'activité ovarienne saison-niére chez la brebis. *Proc. VIᵉ Congr. int. Reprod. anim. Insem. artif., Paris.*

220. LESLIE, P. H., PERRY, J. S. and WATSON, J. S. 1945. The determination of the median body-weight at which female rats reach maturity. *Proc. zool. Soc. Lond.* **115**, 473.

*221. LIGGINS, G. C. 1968. Premature parturition after infusion of cortico-trophin or cortisol into foetal lambs. *J. Endocr.* **42**, 323.

222. — KENNEDY, P. C. and HOLM, L. W. 1967. Failure of initiation of par-turition after electro-coagulation of the pituitary of the foetal lamb. *Am. J. Obstet. Gynec.* **98**, 1080.

*223. LILLIE, F. R. 1917. The freemartin: a study of the action of sex hormones in the foetal life of cattle. *J. exp. Zool.* **23**, 371.

224. LINCOLN, G. A., YOUNGSON, R. W. and SHORT, R. V. 1970. The social and sexual behaviour of the red deer stag. *J. Reprod. Fert.* Suppl. 11, 71.

225. LINZELL, J. L. and HEAP, R. B. 1968. Comparison of progesterone meta-bolism in the pregnant sheep and goat: sources of production and an estimate of uptake by some target organs. *J. Endocr.* **41**, 433.

226. LIPSCHÜTZ, A. 1928. New developments in ovarian dynamics and the law of follicular constancy. *J. exp. Biol.* **5**, 283.

227. LOEB, L. 1923. The effect of extirpation of the uterus on the life and function of the corpus luteum in the guinea-pig. *Proc. Soc. exp. Biol. Med.* **20**, 441.

*228. LONG, J. A. and EVANS, H. M. 1922. The oestrous cycle in the rat and its associated phenomena. *Mem. Univ. Calif.* **6**, 148.

229. LOSTROH, A. J. 1966. Amounts of interstitial cell-stimulating hormone and follicle-stimulating hormone required for follicular development, uterine growth and ovulation in the hypophysectomised rat. *Endocrinology*, **79**, 991.

230. LUTWAK-MANN, CECILIA, HAY, MARY F. and ADAMS, C. E. 1962. The effect of ovariectomy on rabbit blastocysts. *J. Endocr.* **24,** 185.

231. LYONS, W. R. 1943. Pregnancy maintenance in hypophysectomised-oophorectomised rats injected with estrone and progesterone. *Proc. Soc. exp. Biol. Med.* **54,** 65.

*232. McCANN, S. M. 1968. Present status of hypothalamic hypophyseal stimulating and inhibiting hormones. In: *Proc. 24th Int. Congr. physiol. Sci.* (*Abstr.*), Vol. 6, p. 202. Int. Union physiol. Socs., Washington, D.C.

*233. — DHARIWAL, A. P. S. and PORTER, J. C. 1968. Regulation of the adenohypophysis. *Ann. Rev. Physiol.* **30,** 589.

*234. — and PORTER, J. C. 1969. Hypothalamic pituitary stimulating and inhibiting hormones. *Physiol. Rev.* **49,** 240.

235. McCLURE, T. J. 1967. Infertility in mice caused by fasting at about the time of mating. II. Pathological changes. *J. Reprod. Fert.* **13,** 387.

236. McKEEVER, S. 1966. Reproduction in *Citellus beldingi* and *Citellus lateralis* in northeastern California. In: *Comparative biology of reproduction in mammals.* Ed. I. W. Rowlands, p. 365. Academic Press, New York.

237. McKENZIE, F. F. 1940. Recent reproduction studies on equines. *Rec. Proc. Am. Soc. Anim. Prod.*, p. 98.

238. McLAREN, ANNE 1967. Delayed loss of the zona pellucida from blastocysts of suckling mice. *J. Reprod. Fert.* **14,** 159.

239. McNALTY, A. S. (Ed.) 1965. Butterworths Medical Dictionary. Butterworths, London.

240. MANDL, ANITA M. 1951. The phases of the oestrous cycle in the adult white rat. *J. exp. Biol.* **28,** 576.

241. — and ZUCKERMAN, S. 1951. Numbers of normal and atretic oocytes in unilaterally spayed rats. *J. Endocr.* **7,** 112.

242. MANLEY, G. H. 1966. Reproduction in lorisoid primates. *Symp. Zool. Soc. Lond.* **15,** 493.

243. MARSDEN, H. M. and BROUSON, F. H. 1965. The synchrony of oestrus in mice: relative roles of the male and female environments. *J. Endocr.* **32,** 313.

244. MARSHALL, F. H. A. 1904. The oestrous cycle in the common ferret. *Q. Jl microsc. Sci.* **48,** 323.

*245. — 1936. The Croonian Lecture; sexual periodicity and the causes which determine it. *Phil. Trans. R. Soc. Ser. B.* **226,** 423.

246. — 1942. Exteroceptive factors in sexual periodicity. *Biol. Rev.* **17,** 68.

247. MARTÍNEZ-ESTEVE, P. 1937. Le cycle sexuel vaginal chez le marsupial *Didelphys azarae. C.r. Séanc. Soc. Biol.* **124,** 502.

*248. MARTINI, L., CARRARO, A., CAVIEZEL, F. and FOCHI, M. 1968. Factors affecting hypothalamic functions: the pharmacology of puberty. In: *Pharmacology of reproduction.* Ed. E. Diczfalusy. Pergamon Press, Oxford.

249. — FRASCHINI, F. and MOTTA, M. 1967. Hypothalamic mechanisms and anterior pituitary functions. *Schering symposium on endocrinology,* Berlin, p. 201. Ed. G. Raspe. Pergamon Press, Oxford.

*250. — FRASCHINI, F. and MOTTA, M. 1968. Neural control of anterior pituitary functions. In: *Recent progress in hormone research.* Vol. 24. Ed. E. B. Astwood, p. 439. Academic Press, New York.

251. MASUDA, H., ANDERSON, L. L., HENRICKS, D. M. and MELAMPY, R.M. 1967. Progesterone in ovarian venous plasma and corpora lutea of the pig. *Endocrinology* 80, 240.

*252. MATTHEWS, L. H. 1952. *British mammals.* 'New Naturalist' series. Collins, London.

*253. MAYER, G. 1963. Delayed nidation in rats: a method of exploring the mechanisms of ovo-implantation. In: *Delayed implantation.* Ed. A. C. Enders. University of Chicago Press.

254. MICHAEL, R. P. 1965. Oestrogens in the central nervous system. *Br. med. Bull.* 21, 87.

255. — 1968. Gonadal hormones and the control of primate behaviour. In: *Endocrinology and human behaviour.* Ed. R. P. Michael. Oxford University Press, London.

256. MIKHAIL, G., NOALL, M. W. and ALLEN, W. M. 1961. Progesterone levels in the rabbit ovarian vein blood throughout pregnancy. *Endocrinology* 69, 504.

*257. MOOR, R. M. and ROWSON, L. E. A. 1966a. Local uterine mechanisms affecting luteal function in the sheep. *J. Reprod. Fert.* 11, 307.

258. — and ROWSON, L. E. A. 1966b. The corpus luteum of the sheep: functional relationship between the embryo and the corpus luteum. *J. Endocr.* 34, 233.

259. — and ROWSON, L. E. A. 1966c. The corpus luteum of the sheep: effect of the removal of embryos on luteal function. *J. Endocr.* 34, 497.

260. — and ROWSON, L. E. A. 1966d. Local maintenance of the corpus luteum in sheep with embryos transferred to various isolated portions of the uterus. *J. Reprod. Fert.* 12, 539.

261. MOORE, W. W. and NALBANDOV, A. V. 1953. Neurogenic effects of uterine distension on the estrous cycle of the ewe. *Endocrinology* 53, 1.

262. MORRIS, B. 1961. Some observations on the breeding season of the hedgehog and the rearing and handling of the young. *Proc. zool. Soc. Lond.* 136, 201.

263. MORRIS, B. and SASS, M. B. 1966. The formation of lymph in the ovary. *Proc. R. Soc. B.* 164, 577.

264. MOSSMAN, H. W. 1937. The thecal gland and its relation to the reproductive cycle. A study of the cyclic changes in the ovary of the pocket
gopher (Geomys bursarius Shaw). Am. J. Anat. 61, 289.

265. — and JUDAS, I. 1949. Accessory corpora lutea, lutein cell origin, and the
ovarian cycle in the Canadian porcupine. Am. J. Anat. 85, 1.

*266. MOUSTGAARD, J. 1959. Nutrition and reproduction in domestic animals.
In: Reproduction in domestic animals. Ed. H. H. Cole and P. T. Cupps.
Academic Press, New York.

267. MYERS, K. and POOLE, W. E. 1962. A study of the biology of the wild
rabbit, Oryctolagus cuniculus (L.), in confined populations. III. Reproduction. Aust. J. Zool. 10, 225.

*268. NALBANDOV, A. V. 1961. Mechanisms controlling ovulation of avian and
mammalian follicles. In: Control of ovulation, p. 122. Ed. C. A. Villee,
Pergamon Press, Oxford.

269. — MOORE, W. W. and NORTON, H. W. 1955. Further studies on the
neurogenic control of the estrous cycle by uterine distension. Endocrinology 56, 225.

270. NAWITO, M. F., SHALASH, M. R., HOPPE, R. and RAKHA, A. M. 1967.
Reproduction in female camel. Bull. Anim. Sci. Res. Inst., Cairo 2.

271. NEAL, E. and HARRISON, R. J. 1958. Reproduction in the European
badger. Trans. zool. Soc. Lond. 29, 67.

272. NELSON, W. W. and GREENE, R. R. 1953. The human ovary in pregnancy.
Int. Abstr. Surg. 97, 1.

273. NEUMANN, F., ELGER, W. and STEINBECK, H. 1969. Drug-induced intersexuality in mammals. J. Reprod. Fert., Suppl. 7, 9.

274. NEWSOME, A. E. 1965. Reproduction in natural populations of the red
kangaroo, Megaleia rufa (Desmarest) in central Australia. Aust. J. Zool.
13, 735.

275. — 1966. The influence of food on breeding in the red kangaroo in central
Australia. C.S.I.R.O. Wildl. Res. 11, 187.

276. NEWSON, R. M. 1966. Reproduction in the feral coypu (Myocastor
coypus). In: Comparative biology of reproduction in mammals. Ed. I. W.
Rowlands, p. 323. Academic Press, New York.

277. O'DONOGHUE, P. N. 1963. Reproduction in the female hyrax. Proc. zool.
Soc. Lond. 141, 207.

278. ORSINI, MARGARET W. and McLAREN, ANNE. 1967. Loss of the zona
pellucida in mice, and the effect of tubal ligation and ovariectomy.
J. Reprod. Fert. 13, 485.

279. OTA, M. and HSIEH, K. S. 1968. Failure of melatonin to inhibit ovulation
induced with pregnant mare serum and human chorionic gonadotrophins in rats. J. Endocr. 41, 601.

280. PARKES, A. S. 1926. On the occurrence of the oestrous cycle after X-ray
sterilization. I. Irradiation of mice at three weeks old. Proc. R. Soc.
B. 100, 172.

281. — 1928. The length of the oestrous cycle in the unmated normal mouse: records of one thousand cycles. *Brit. J. exp. Biol.* **5**, 371.

282. — 1931. The reproductive processes of certain mammals. Part 1. The oestrous cycle of the Chinese hamster (*Cricetulus griseus*). *Proc. R. Soc. B.* **108**, 138.

*283. — 1966. Sex, Science and Society. Oriel Press, Newcastle-upon-Tyne.

284. — and BELLERBY, C. W. 1927. Studies on the internal secretions of the ovary. III. The effects of injection of oestrin during lactation. *J. Physiol. Lond.* **62**, 301.

*285. — and DEANESLY, R. 1966. The ovarian hormones. In: *Marshall's Physiology of reproduction.* 3rd ed. Ed. A. S. Parkes. Vol. 3, p. 570. Longmans, London.

*286. PARLOW, A. F. 1961. Bio-assay of pituitary luteinizing hormone by depletion of ovarian ascorbic acid. In: *Human pituitary gonadotropins*, p. 300. Ed. A. Albert. C. C. Thomas, Springfield, Illinois.

287. PARR, E. L., SCHAEDLER, R. W. and HIRSCH, J. G. 1967. The relationship of polymorphonuclear leukocytes to infertility in uteri containing foreign bodies. *J. exp. Med.* **126**, 523.

288. PEARSON, O. P. 1944. Reproduction in the shrew (*Blarina brevicauda*, Say). *Am. J. Anat.* **75**, 39.

289. — 1949. Reproduction of a South American rodent, the mountain viscacha. *Am. J. Anat.* **84**, 143.

290. PERRY, J. S. 1945. The reproduction of the wild brown rat (*Rattus norvegicus* Erxleben). *Proc. zool. Soc. Lond.* **115**, 19.

291. — 1953. The reproduction of the African elephant, *Loxodonta africana*. *Phil. Trans. R. Soc.*, Ser. B. **237**, 93.

*292. — (Ed.) 1969. Intersexuality. *J. Reprod. Fert.*, Suppl. 7.

*293. — and ROWLANDS, I. W. 1962. The ovarian cycle in vertebrates. In: *The ovary*, p. 275. Ed. S. ZUCKERMAN. Academic Press, New York.

294. — and ROWLANDS, I. W. 1962. Early pregnancy in the pig. *J. Reprod. Fert.* **4**, 175.

295. PETERS, H., LEVY, E. and CRONE, M. 1965. Oogenesis in rabbits. *J. exp. Zool.* **158**, 169.

296. PETRUSEWICZ, K. 1958. Investigation of experimentally induced population growth. *Ekol. Pol.* **5**, 281.

297. PFEIFFER, C. A. 1935. Origin of functional differences between male and female hypophyses. *Proc. Soc. exp. Biol. Med.* **32**, 603.

*298. — 1936. Sexual differences of the hypophyses and their determination by the gonads. *Am. J. Anat.* **58**, 195.

299. PICON, L. 1956. Sur le rôle de l'age dans la sensibilité de l'ovaire à l'hormone gonadotrope chez le rat. *Archs Anat. microsc. Morph. exp.* **45**, 311.

300. PIMLOTT, D. H. 1959. Reproduction and productivity of Newfoundland moose. *J. Wildl. Mgmt* **23**, 381.

301. PLOTKA, E. D., ESTERGREEN, V. L. and FROST, O. L. 1967. Relationships between progesterone levels in peripheral and ovarian venous blood plasma, corpora lutea and ovaries during pregnancy. *J. Dairy Sci.* **50**, 1001.

302. POLGE, C., DAY, B.N. and GROVES, T. W. 1968. Synchronisation of ovulation and artificial insemination in pigs. *Vet. Rec.* **83**, 136.

303. POOLE, W. E. and PILTON, PHYLLIS E. 1964. Reproduction in the grey kangaroo, *Macropus canguru*, in captivity. *C.S.I.R.O. Wildlife Research* **9**, 218.

*304. POPA, G. T. and FIELDING, UNA 1930. A portal circulation from the pituitary to the hypothalamic region. *J. Anat.* **65**, 88.

*305. PORTER, D. G. 1969. Progesterone and the guinea-pig myometrium. In: *Progesterone: its regulatory effect on the myometrium.* (Ciba Foundation Study Group No. 34.) Ed. G. E. W. Wolstenholme and Julie Knight. p. 79. Churchill, London.

306. PRICE, DOROTHY, ORTIZ, EVELINA and ZAAIJER, JOHANNA J. P. 1967. Organ culture studies of hormone secretion in endocrine glands of fetal guinea-pigs. III. The relation of testicular hormone to sex differentiation of the reproductive ducts. *Anat. Rec.* **157**, 27.

307. PRICE, M. 1953. The reproductive cycle of the water shrew, *Neomys fodiens bicolor* Shaw. *Proc. zool. Soc. Lond.* **123**, 599.

308. RAESIDE, J. I. and TURNER, C. W. 1955. Chemical estimation of progesterone in the blood of cattle, sheep and goats. *J. Diary Sci.* **38**, 1334.

309. RAO, C. R. N. 1932. On the occurrence of glycogen and fat in liquor folliculi and uterine secretion in *Loris lydekkerianus* Cabr. *J. Mysore Univ.* **6**, 140.

310. REITER, R. J. 1968. Changes in the reproductive organs of cold-exposed and light-deprived female hamsters (*Mesocricetus auratus*). *J. Reprod. Fert.* **16**, 217.

311. RICHMOND, M. and CONAWAY, C. H. 1969. Induced ovulation and oestrus in *Microtus ochrogaster*. *J. Reprod. Fert.* Suppl. 6, 357.

*312. RIDDLE, O. 1963. Prolactin in vertebrate function and organization. *J. natn. Cancer Inst.* **31**, 1039.

313. ROBINSON, T. J. 1952. The role of progesterone in the mating behaviour of the ewe. *Nature, Lond.* **170**, 373.

*314. ROBSON, J. M. 1937. Maintenance by oestrin of the luteal function in hypophysectomized rabbits. *J. Physiol.* **90**, 435.

315. ROSENBUSCH-WEIHS, D. and PONSE, K. 1957. Actions rapide et lointaines de l'hypophysectomie chez le cobaye. *Rev. Suisse Zool.* **64**, 271.

*316. ROTHCHILD, I. 1966. The nature of the luteotrophic process. *J. Reprod. Fert.* Suppl. 1, 49.

*317. ROWAN, W. 1938. Light and seasonal reproduction in animals. *Biol. Rev.* **12**, 374.

318. ROWLANDS, I. W. 1949. Serum gonadotrophin and ovarian activity in the pregnant mare. *J. Endocr.* **6,** 184.

319. — 1956. The corpus luteum of the guinea-pig. *Ciba Foundation Colloquia on Ageing* **2,** 69.

320. — 1962. Effect of oestrogens, prolactin and hypophysectomy on the corpora lutea and vagina of hysterectomized guinea-pigs. *J. Endocr.* **24,** 105.

321. — and HEAP, R. B. 1966. Histological observations on the ovary and progesterone levels in the coypu, *Myocastor coypus.* In: *Comparative biology of reproduction in mammals.* Ed. I. W. Rowlands, p. 335. Academic Press, New York.

322. — and SHORT, R. V. 1959. The progesterone content of the guinea-pig corpus luteum during the reproductive cycle and after hysterectomy. *J. Endocr.* **19,** 81.

323. ROWSON, L. E. A. 1951. Methods of inducing multiple ovulation in cattle. *J. Endocr.* **7,** 260.

324. — and MOOR, R. M. 1966. Embryo transfer in the sheep: the significance of synchronizing oestrus in the donor and recipient animal. *J. Reprod. Fert.* **11,** 207.

325. — and MOOR, R. M. 1967. The influence of embryonic tissue homogenate, infused into the uterus, on the life-span of the corpus luteum in the sheep. *J. Reprod. Fert.* **13,** 511.

326. RUBIN, B. L., DEANE, H. W. and HAMILTON, J. A. 1963. Biochemical and histochemical studies on Δ^5-3β-hydroxysteroid dehydrogenase activity in the adrenal glands and ovaries of diverse mammals. *Endocrinology* **73,** 748.

327. — HILLIARD, J., HAYWARD, J. N. and DEANE, H. W. 1965. Acute effects of gonadotrophic hormones on rat and rabbit ovarian Δ^5-3β-hydroxysteroid dehydrogenase activities. *Steroids.* Suppl. 1, 121.

328. RUBY, J. R., DYER, R. F. and SKALKO, R. G. 1969. The occurrence of intercellular bridges during oogenesis in the mouse. *J. Morph.* **127,** 307.

*329. SADLEIR, R. M. F. S. 1969a. The role of nutrition in the reproduction of wild mammals. *J. Reprod. Fert.*, Suppl. 6, 39.

*330. — 1969b. *The ecology of reproduction in wild and domestic mammals.* Methuen, London.

331. SALE, J. B. 1969. Breeding season and litter size in Hyracoidea. *J. Reprod. Fert.*, Suppl. 6, 249.

332. SANDBERG, A. A., ROSENTHAL, H., SCHNEIDER, S. L. and SLAUNWHITE, W. R., Jr. 1966. Protein-steroid interactions and their role in the transport and metabolism of steroids. In: *Steroid dynamics.* Ed. G. Pincus, T. Nakao and J. F. Tait, p. 1. Academic Press, New York.

333. SAVARD, K., MARSH, J. M. and RICE, B. F. 1965. Gonadotropins and ovarian steroidogenesis. *Rec. Prog. Horm. Res.* **21,** 285.

334. SCHILLER, E. L. 1956. Ecology and health of *Rattus* at Nome, Alaska. *J. Mammal.* **37,** 181.

335. SCHINCKEL, P. G. 1954. The effect of the ram on the incidence and occurrence of oestrus in ewes. *Aust. vet. J.* **30,** 189.

336. — 1954. The effect of the presence of the ram on the ovarian activity of the ewe. *Aust. J. agric. Res.* **5,** 465.

337. SCHMIDT-ELMENDORFF, H., LORAINE, J. A. and BELL, E. T. 1962. The effect of 6M urea on the follicle-stimulating and luteinizing hormone activities of various gonadotrophin preparations. *J. Endocr.* **24,** 153.

338. SCHOMBERG, D. W. 1967. Demonstration *in vitro* of luteolytic activity in pig uterine flushings. *J. Endocr.* **38,** 359.

339. SCHWARTZ, N. B. and CALDARELLI, D. 1965. Plasma LH in cyclic female rats. *Proc. Soc. exp. Biol. Med.* **119,** 16.

*340. — and HOFFMANN, J. C. 1967. Model for the control of the mammalian reproductive cycle. In: *Proc. 2nd Int. Congr. Hormonal Steroids* (Milan, May 1966). Ed. L. Martini and F. Fraschini.

341. SCOTT, P. P. and LLOYD-JACOB, M. A. 1955. Some interesting features in the reproductive cycle of the cat. *Stud. Fertil.* **7,** 123.

*342. SEAL, U. S. and DOE, R. P. 1966. Corticosteroid-binding globulin: biochemistry, physiology and phylogeny. In: *Steroid dynamics.* Ed. G. Pincus, T. Nakao and J. F. Tait, p. 63. Academic Press, New York.

*343. SELYE, H. 1946. The general adaptation syndrome and the diseases of adaptation. *J. clin. Endocr.* **6,** 117.

*344. SHARMAN, G. B. 1963. Delayed implantation in marsupials. In: *Delayed implantation.* Ed. A. C. Enders. University of Chicago Press.

345. — and PILTON, PHYLLIS E. 1964. The life-history and reproduction of the red kangaroo (*Megaleia rufa*). *Proc. zool. Soc. Lond.* **142,** 29.

*346. SHELASNYAK, M. C. and KRAICER, P. F. 1963. The role of oestrogen in nidation. In: *Delayed implantation.* Ed. A. C. Enders. University of Chicago Press.

347. SHEPPE, W. A. 1965. Unseasonal breeding in artificial colonies of *Peromyscus leucopus. J. Mammal.* **46,** 641.

348. SHETTLES, L. B. 1957. The living human ovum. *Obstet. Gynec.* **10,** 359.

349. SHORT, R. V. 1959. Progesterone in blood. IV. Progesterone in the blood of mares. *J. Endocr.* **19,** 207.

350. — 1961. Progesterone. In: *Hormones in blood.* Ed. C. H. Gray and A. L. Bacharach, p. 379. Academic Press, New York.

351. — 1962. Steroids in the follicular fluid and the corpus luteum of the mare. A 'two-cell type' theory of ovarian steroid synthesis. *J. Endocr.* **24,** 59.

352. — 1964. Ovarian steroid synthesis and secretion *in vivo. Recent Prog. Horm. Res.* **20,** 303.

353. — 1966. Oestrous behaviour, ovulation and the formation of the corpus luteum in the African elephant, *Loxodonta africana. Afr. Wildl. J.* **4,** 56.

*354. — 1967. Reproduction. *A. Rev. Physiol.* **29**, 373.

*355. — 1969. Implantation and the maternal recognition of pregnancy. In: *Foetal autonomy.* Ed. G. E. W. Wolstenholme and M. O'Connor, p. 2. Churchill, London.

356. — and HAY, MARY F. 1965. Delayed implantation in the roe deer. *J. Reprod. Fert.* **9**, 372.

357. SINGH, K. B. and GREENWALD, G. S. 1967. Effects of continuous light on the reproductive cycle of the female rat: induction of ovulation and pituitary gonadotrophins during persistent oestrus. *J. Endocr.* **38**, 389.

358. SKINNER, J. D. 1967. Puberty in the male rabbit. *J. Reprod. Fert.* **14**, 151.

359. SMELSER, G. K., WALTON, A. and WHETHAM, E. O. 1934. The effect of light on ovarian activity in the rabbit. *J. exp. Biol.* **11**, 352.

360. SMITH, B. D. and BRADBURY, J. T. 1963. Ovarian response to gonadotrophins after pre-treatment with diethylstilbestrol. *Am. J. Physiol.* **204**, 1023.

361. — and BRADBURY, J. T. 1966. Influence of progestins on ovarian responses to estrogen and gonadotrophins in immature rats. *Endocrinology* **78**, 297.

362. SMITH, I. D. 1965. The influence of level of nutrition during winter and spring upon oestrous activity in the ewe. *Wld Rev. Anim. Prod.* **4**, 95.

*363. SMITH, P. E. and ENGLE, E. T. 1927. Experimental evidence regarding the role of the anterior pituitary in the development and regulation of the genital system. *Am. J. Anat.* **40**, 159.

364. SMITH, R. E. and FARQUHAR, MARILYN G. 1966. Lysosome function in the regulation of the secretory process in cells of the anterior pituitary gland. *J. Cell Biol.* **31**, 319.

365. SMITH, V. R., McSHAN, W. H. and CASIDA, L. E. 1957. On maintenance of the corpora lutea of the bovine with lactogen. *J. Dairy Sci.* **40**, 443.

366. SORENSON, M. W. and CONAWAY, C. H. 1968. The social and reproductive behaviour of *Tupaia montana* in captivity. *J. Mammal.* **49**, 502.

367. STAFFORD, W. T., COLLINS, R. F. and MOSSMAN, H. W. 1942. The thecal gland in the guinea-pig ovary. *Anat. Rec.* **83**, 193.

*368. STOCKARD, C. R. and PAPANICOLAOU, G. N. 1917. The existence of a typical oestrous cycle in the guinea-pig—with a study of its histological and physiological changes. *Am. J. Anat.* **22**, 225.

369. STUKOVSKY, R. and VALŠÍK, J. A. 1966. Mois de naissance et puberté chez les filles. *Biometrie humaine* **1**, 25.

370. SUTHERLAND, E. W. and RALL, T. W. 1960. The relation of adenosine-3′,5′-phosphate and phosphorylase to the actions of catecholamines and other hormones. *Pharmac. Rev.* **12**, 265.

*371. TANNER, J. M. 1967. Puberty. In: *Advances in reproductive physiology*, Vol. 2. Ed. Anne McLaren. Academic Press, New York.

372. TURBERVILE, G. 1576. *The noble arte of venerie or hunting.* Clarendon Press, Oxford, 1908.

373. TURNER, C. D. 1966. *General endocrinology*. 4th edition. W. B. Saunders, Philadelphia.

*374. TYNDALE-BISCOE, C. H. 1963. The role of the corpus luteum in the delayed implantation of marsupials. In: *Delayed implantation*. Ed. A. C. Enders. University of Chicago Press.

375. UNDERWOOD, T. 1969. The cell's 'messenger boy'. *New Scientist* 41, 692.

376. VANDENBERGH, J. G. 1967. Male influence on the sexual development of female mice. *Gen. comp. Endocr.* 9, 522.

377. VAN DER HORST, C. J. 1954. *Elephantulus* going into anoestrus; menstruation and abortion. *Phil. Trans. R. Soc. Ser. B.* 238, 27.

378. — and GILLMAN, J. 1940. Ovulation and corpus luteum formation in *Elephantulus*. *S. Afr. J. med. Sci.* 5, 73.

379. — and GILLMAN, J. 1941. The menstrual cycle in *Elephantulus*. *S. Afr. J. med. Sci.* 6, 27.

380. — and GILLMAN, J. 1946. The corpus luteum of *Elephantulus* during pregnancy; its form and function. *S. Afr. J. med. Sci.* 11, Biol. Suppl. 87.

*381. VAN DER LEE, S. and BOOT, L. M. 1955. Spontaneous pseudopregnancy in mice. *Acta physiol. pharmac. néerl.* 4, 442.

382. VANDE WIELE, R. L. and TURKSOY, R. N. 1965. Treatment of amenorrhea and of anovulation with human menopausal and chorionic gonadotropins. *J. clin. Endocr. Metab.* 25, 369.

383. VAN LANCKER, J. L. 1964. Lysosomes. *Fed. Proc.* 23, 1009.

384. VARAVUDHI, P. 1968. Induction of nidation in hypophysectomized progesterone-treated rats by non-specific factors. *J. Endocr.* 40, 429.

385. WATSON, J. S. and PERRY, J. S. 1954. Experiments on rat control in Palestine and the Sudan. In: *Control of rats and mice*, Vol. 2, 500. Ed. D. Chitty. Clarendon Press, Oxford.

386. WATSON, R. M. 1969. Reproduction of wildebeest (*Connochaetes taurinus albojubatus* Thomas) in the Serengeti region, and its significance to conservation. *J. Reprod. Fert.* Suppl. 6, 289.

387. WEIR, BARBARA J. Reproduction of the plains viscacha. *J. Reprod. Fert.* 20, 358.

*388. WHITTEN, W. K. 1956. Modification of the oestrous cycle of the mouse by external stimuli associated with the male. *J. Endocr.* 13, 399.

389. — 1959. Occurrence of anoestrus in mice caged in groups. *J. Endocr.* 18, 102.

390. WIMSATT, W. A. and PARKS, H. F. 1966. Ultrastructure of the surviving follicle of hibernation and of the ovum-follicle cell relationship in the vespertilionid bat, *Myotis lucifugus*. In: *Comparative biology of reproduction in mammals*. Ed. I. W. Rowlands, p. 419. Academic Press, New York.

391. WINIWARTER, H. DE 1901. Recherches sur l'ovogenèse et l'organogenèse de l'ovaire des mammifères (lapin et homme). *Archs Biol., Liège* 17, 33.

210   THE OVARIAN CYCLE OF MAMMALS

392. WINTENBERGER-TORRES, S. and ROMBAUTS, P. 1968. Relation entre la mortalité embryonnaire et la quantité de progestérone secretée chez la brebis. *VIe Congr. Reprod. Insem. Artif., Paris* **1**, 491.

393. WISLOCKI, G. B. 1931. Notes on the female reproductive tract (ovaries, uterus and placenta) of the collared peccary (*Pecari angulatus* Bangsi Goldman). *J. Mammal.* **12**, 143.

394. — and STREETER, G. L. 1938. On the placentation of the macaque (*Macaca mulatta*), from the time of implantation until the formation of the definitive placenta. *Contr. Embryol.* **27**, 1.

*395. WURTMAN, R. J. 1969. The pineal gland in relation to reproduction. *Am. J. Obstet. Gynec.* **104**, 320.

*396. — AXELROD, J. and KELLY, D. E. 1968. *The pineal.* Academic Press, New York.

397. ZAMBONI, L. and MASTROIANNI, L., Jr. 1966. Electron microscope studies on rabbit ova. 1. The follicular oocyte. *J. ultrastruct. Res.* **14**, 95.

398. ZONDEK, B. and ASCHHEIM, S. 1927. Hypophysenvorderlappen und Ovarium. Beziehungen der endokrinen Drüsen zur Ovarialfunktion. *Arch. Gynäk* **130**, 1.

Index

211

Vitelline membrane, 23
Voles: exteroceptive factors in breeding, 163, 183; induced ovulation, 78–9; pregnancy concurrent with lactation, 119; puberty, 153

Weasel, 78
Whitten effect, 177, 179 *et seq.*
Wildebeeste (synchrony of oestrus in wild), 182

Wolffian duct (see Differentiation of sexes)

Y chromosome, 8, 11, 66

Zebra, 81
Zona pellucida: formation and function, 23–4; at implantation, 121; ultrastructure, 35–6